Introduction to Video Search Engines

David C. Gibbon • Zhu Liu

Introduction to Video Search Engines

Springer

David C. Gibbon
AT & T Labs Research
200 Laurel Ave.
Middletown NJ 07748
USA
dcg@research.att.com

Zhu Liu
AT & T Labs Research
200 Laurel Ave.
Middletown NJ 07748
USA
zliu@research.att.com

ISBN 978-3-540-79336-6 ISBN 978-3-540-79337-3 (eBook)

Library of Congress Control Number: 2008932565

ACM Computing Classification (1998): H.2, H.3, H.5, 1.2.10, 1.4

Cover design: KünkelLopka, Heidelberg

9 8 7 6 5 4 3 2 1

springer.com

Preface

The evolution of technology has set the stage for the rapid growth of the video Web: broadband Internet access is ubiquitous, and streaming media protocols, systems, and encoding standards are mature. In addition to Web video delivery, users can easily contribute content captured on low cost camera phones and other consumer products. The media and entertainment industry no longer views these developments as a threat to their established business practices, but as an opportunity to provide services for more viewers in a wider range of consumption contexts. The emergence of IPTV and mobile video services offers unprecedented access to an ever growing number of broadcast channels and provides the flexibility to deliver new, more personalized video services. Highly capable portable media players allow us to take this personalized content with us, and to consume it even in places where the network does not reach.

Video search engines enable users to take advantage of these emerging video resources for a wide variety of applications including entertainment, education and communications. However, the task of information extraction from video for retrieval applications is challenging, providing opportunities for innovation. This book aims to first describe the current state of video search engine technology and second to inform those with the requisite technical skills of the opportunities to contribute to the development of this field.

Today's Web search engines have greatly improved the accessibility and therefore the value of the Web. The top portals prominently feature search capabilities and go beyond text search to include image and video search in various forms. A number of smaller companies have begun to offer more sophisticated media search features based on content analysis. Academic research groups have been actively developing algorithms and prototypes in this area for over a decade; incorporating and advancing previously existing constituent technologies.

Most media search systems rely on available metadata or contextual information in text form. Syndication formats such as RSS provide organized access to media sources and include descriptive global metadata. While these information sources are valuable and should be exploited, they are limited because they are typically brief, high level and subjective.

Therefore the current focus of media indexing research is to develop algorithms to exploit the media content itself as much as possible to augment available metadata. In some cases, the media may contain associated text streams such as closed caption or song lyrics. By extracting and operating on these streams, a textual representation of the dialog is obtained and existing text information retrieval methods can then be applied to retrieve relevant media. Speech recognition can be employed to create an approximation of the transcription, and techniques such as video optical character recognition can also be used to generate a textual representation of the media content. Although these technologies are inherently error prone, they have been used with success for indexing applications. Advanced speech retrieval systems use phonetic search to deal with the "out of vocabulary" problem and maintain alternative hypotheses in the form of lattices to boost recall.

Media retrieval that goes beyond the textual media component is more complex because the basic media features are not well defined and may not scale well for large archives. Further, formulating queries may not be as simple as typing a keyword. However, systems have been designed to, for example, retrieve images similar to a given image (query by example) or retrieve images based on a specification of color or shape. For navigating video retrieval results, techniques such as video skimming or mosaicing have been proposed.

The book will have a practical emphasis with the goal of bringing researchers up to date on the state of the art in multimedia search technologies and systems. Part of the presentation will follow a logical flow from content acquisition, analysis to extract index data, data representation, media archival, retrieval and finally rendering results in a Web-based environment. Each of these major functional components will be outlined, and particular emphasis will be given to automated content analysis techniques since this is critical for operating video search engines at scale, and it presents on-going research challenges. To give the readers an understanding of the issues involved, individual media processing algorithms operating on text, audio and video will be addressed including text alignment, case restoration, entity extraction, speech recognition, speaker segmentation, and video shot boundary detection. Additionally, the value of operating on multiple media components simultaneously will be illustrated by examining multimodal processing techniques, e.g. for media segmentation. The role of media segmentation in improving relevance ranking for long-form content will be discussed.

Who should read this book?

The book is intended for senior undergraduates or first-year graduate students in computer science or computer engineering, as well as professionals working in related fields. Although not intended for experts working directly on video search engines, the book will present a refreshing, broad perspective on video search and will have value as a reference tool. The topic of multimedia search spans multiple disciplines so the book will be valuable to experts in the constituent technologies such as speech processing or information retrieval who are looking to broaden their knowledge beyond their current areas of expertise.

A basic knowledge of Web application technologies, databases and computer networking issues is assumed. While a basic knowledge of the constituent technologies would be helpful, the intent will be to present these at an introductory level and discuss only the elements applicable to the problem of video search. The book explains the overall process of video content acquisition, indexing and retrieval with browsing, provides overviews of constituent technologies such as information retrieval, Internet video systems, video and multimedia processing to extract index data, and gives examples of existing systems and describes their features. The readers will:

- understand at a basic level all of the technologies used in today's video search engines;
- learn which video indexing techniques are appropriate for a given type of video material and be able to make inferences about which methods will work for new video content types;
- be able to differentiate between proven, practical techniques and those that are speculative, under development, or of narrow applicability;
- be able to determine which topics in video search are of interest to them for further study.

How is this book organized?

The book is divided into three main sections:

I. Background and fundamentals: Chapter 1 outlines the technology trends which dictate that video material will increasingly be made available on the Web and points out the challenges that video is much more difficult to search than text files, and it is more difficult to browse. Chapter 2 addresses the nature, availability, and attributes of different sources of video data. Details about available metadata for different types of video (e.g. electronic program guides, transcripts, etc.) are also provided. Chapter 3 reviews Internet video systems, including topics such as bandwidth, compression, random access, streaming, standards, digital rights management, redirector files, etc. Chapter 4 introduces video search engine systems: the process of content acquisition, media processing, building a multimedia database, retrieval, media browsing.

II. Media processing: To address the challenges, we need to move beyond existing metadata retrieval systems, and analyze the content to extract information for indexing. Chapter 5 gives an overview on automated methods, systems, and algorithms for processing media to extract information for indexing and retrieval purposes. Chapters 6 - 8 discuss the specific media processing technologies that are developed in video, audio, and text domains. Multimodal processing, which is designed to mitigate the error that is inevitable with single modal processing, is discussed in Chapter 9.

III. Case studies: Chapter 10 reinforces the concepts of video processing through illustrative examples, and provides details about existing solutions. Practical issues are brought to light through presentation of a detailed case study including a system supporting rapid content queries on a 50,000 hour broadcast television archive spanning 10 years and supporting a wide range of streaming media types for different applications. Chapter 11 provides a review of currently deployed Web search engines and identifies a few trends in the field to provide a sense of future directions.

Acknowledgements

This book became possible due to the support and vision of Dr. Behzad Shahraray, the head of the Video and Multimedia Services Research Department at AT&T Labs Research, Dr. Richard Cox and Dr. Lawrence Rabiner, directors of the Speech and Image Processing Research Laboratory, and Prof. Yang Wang at the Electrical Engineering department of Polytechnic University. Our work has been inspired by our colleagues, Lee Begeja, Bernard Renger, Harris Drucker, Andrea Basso and Murat Saraclar. We also benefit from collaborations with Prof. Shi-fu Chang and Eric Zavesky at Columbia University. This book began as a tutorial that the authors gave at the WWW2006 Conference in Edinburgh Scotland as suggested by Robin Chen. All their support and help is greatly appreciated.

Contents

1 Video Search

1.1 Introduction

Today's World Wide Web is truly a video Web. Millions of video clips are available to users instantly thanks to widely available broadband IP networks, low-cost storage and mature digital video delivery technologies. The content of this video runs the gamut from skateboarding antics captured on mobile phone cameras up through graduate level university lecture series on computer science. On the commercial entertainment side, all major broadcasters and movie studios have on-line strategies which range from a focus on promoting traditional distribution channels through releasing primetime programming through the Web to capture an emerging demographic who increasingly turn to their laptops for video entertainment instead of their televisions.

Although Internet video systems have made great strides, television still provides the highest quality digital video available to consumers on a daily basis at a level of quality far beyond that of traditional best-effort Internet video streaming. On-line high definition (HD) content is still a novelty. Cable, direct broadcast satellite, and over-the-air digital broadcast are mature technologies providing HD quality entertainment to millions of consumers today, and Internet protocol television (IPTV) is emerging to provide more functionality and provide increased convergence with existing Web technologies. High capacity digital video recorders allow consumers to easily capture many hours of content for viewing at their convenience. An increasing array of set-top devices and smart TVs with IP connectivity provides access to the wealth of Web video (e.g. expatriates can view news from home in their native langue) via streaming or on demand. Closer to home, consumers can browse their videos and photos captured from their digital cameras or purchased on line and archived on their home network server.

The video Web extends beyond fixed appliances to mobile devices which support fully functional browsers and media players. Battery life and user consumption contexts may limit viewing of long-form content, but specially designed short-form content provides valuable entertainment for opportunistic consumption scenarios such as waiting for a late bus.

This dazzling array of options for access to high quality video anytime, anywhere provides opportunities for service and technology providers who can develop services and tools to help users manage their media experience and locate content of interest from the vast ocean of irrelevant or even repulsive material. Service providers plan to build universal three-screen services to allow users to seamlessly switch from TV to PC to mobile viewing. In fact, three screens are not enough to encompass all of today's consumption scenarios and the term "any screen" is used to include personal media players, portable gaming devices and others such as Internet connected picture frames.

Given the potential impact of technological breakthroughs in video services, it is not surprising that there is no shortage of academic and industrial research groups focused on this task. Further, successful solutions require an interdisciplinary approach, drawing from diverse fields including information retrieval, natural language processing, data mining, machine learning, multimedia databases, as well as speech, audio, image and video processing [Haupt05]. Data visualization, user interface design, human computer interfaces and the consideration of social aspects of media consumption and interaction such as rating, tagging and recommendation are of equal or perhaps greater importance than the media processing technologies. Improvements to the start of the art of video search can draw from a broad base indeed.

1.2 Addressing the Opportunity

Realizing that inexpensive storage, ubiquitous broadband Internet access, low cost digital cameras, and nimble video editing tools would result in a flood of unorganized video content, researchers have been developing video search technologies for a number of years. The recent trends in digital video creation and delivery technology have brought the need for such tools to the forefront. The computing technologies contributing to this flood are also available to the tool builders to help provide a lifeline to Web video viewers. Once-impractical media analysis technologies are being applied to large archives of video content to extract metadata to aid search. The social aspect (e.g. incorporating popularity of pages into rank calculations), initially overlooked and in-fact largely irrelevant due to lack of critical mass, provided a breakthrough for text search engine technology. For video media, the exploitation of user tagging and recommendation engines is similarly providing a much needed boost for video search.

While great advances in video search have been made and today's video search engines provide a valuable service to users, the task of information extraction from video for retrieval applications is challenging; providing opportunities for innovation. This book aims to first describe the current state of video search engine technology and second inform those with the requisite technical skills of the opportunities to contribute to the development of this field.

Today's Web search engines have greatly improved the accessibility and therefore the value of the Web. The top portals prominently feature search capabilities and most have gone beyond text search to include image search and even video search, though the latter on a limited basis. A number of smaller companies have begun to offer more sophisticated media search features. Academic research groups have been actively developing algorithms and prototypes in this area for over a decade; incorporating and advancing previously existing constituent technologies.

Technology evolution has set the stage for rapid growth of video search engines: research and prototyping has been underway for several years, broadband access is ubiquitous, streaming media protocols and encoding standards are mature. Disk and processor cost reductions are making it possible to store and index large volumes of digital media and create indexed on-line archives. Market forces such as the emergence of IPTV and mobile video services and the growing acceptance of digital rights management technologies are fueling these trends.

Most media search systems rely on available metadata or contextual information in text form. Also, surrounding text or anchor text from links to the media are used to infer something about its content and, in some cases RSS feed descriptors point to media and include descriptive metadata. While these information sources are valuable and should be exploited, they are limited because they are typically brief, high level and subjective.

Therefore the current focus of media indexing research is to develop algorithms to exploit the media content itself as much as possible to augment available metadata. In some cases, the media may contain associated text streams such as closed caption or song lyrics. By extracting and operating on these streams, a textual representation of the dialog is obtained and existing text information retrieval methods can then be applied to retrieve relevant media. Speech recognition can be employed to create an approximation of the transcription, and techniques such as video optical character recognition can also be used to generate a textual representation of the media content. Although these technologies are inherently error prone, they have been used with success for indexing applications. Advanced speech retrieval systems use phonetic search to deal with the "out of vocabulary" problem and maintain alternative hypotheses in the form of lattices to boost recall.

Media retrieval that goes beyond the textual media component is more complex because the basic media features are not well defined and may not scale well for large archives. Further, formulating queries may not be as simple as typing a keyword. However systems have been designed to, for example, retrieve images similar to a given image (query by example) or retrieve images based on a specification of color or shape. For navigating video retrieval results, techniques such as video skimming or mosaicing have been proposed.

This book takes a practical approach with the goal of bringing researchers up to date on the state of the art in multimedia search technologies and systems. Part of the presentation will follow a logical flow from content acquisition, analysis to extract index data, data representation, media archival, retrieval and finally rendering results in a Web-based environment. Each of these major functional components will be outlined, and particular emphasis will be given to automated content analysis techniques since this is critical for operating video search engines at scale, and it presents on-going research challenges. To give the readers an understanding of the issues involved, individual media processing algorithms operating on text, audio and video will be addressed including: text alignment, case restoration, entity extraction, speech recognition, speaker segmentation, and video shot boundary detection. Additionally, the value of operating on multiple media components simultaneously will be illustrated by examining multimodal segmentation techniques. The role of media segmentation in improving relevance ranking for long-form content will be discussed.

In addition to media processing, index representation issues using XML and media archival systems will be presented. The relation between indexing, summarization and media adaptation for mobile devices will be discussed. Challenges encountered when building Web-based user interfaces for browsing indexed streaming media will be addressed.

In parallel with the functional discussion, a historical perspective will be provided, and relevant work will be cited from both academic and industrial sources. Background information such as digital media encoding and streaming standards and information retrieval will be given to allow the book to stand on its own.

Application areas vary widely, and the applicability of media search techniques is limited to certain domains. For example, video from Web cams is quite different from broadcast television content. The book will make this clear, pointing out techniques that are suitable for different levels of structure or different quality levels of the source material.

Practical issues will be brought to light through presentation of detailed case studies including a system supporting rapid content queries on a 50,000 hour video archive spanning 10 years of broadcast television and Internet video.

1.3 Classification of Web Video Sites

As users browse the Web, they are likely to encounter video on almost any site. If we focus on the sites that appear to be video portals or claim to offer video search, we can begin to discern several categories of video search sites. To complicate the matter, there are hundreds if not thousands of such sites, and of course the Web is evolving rapidly, with business models and content presentation strategies in a constant state of flux. Some Web destinations are amalgams of several differing approaches. In spite of this, in this section we attempt to point out a few general classes of video Web sites that users may encounter and that employ some form of video search capability.

1.3.1 Content Originators and Traditional Broadcasters

Examples of origin content Web sites include major TV networks NBC.com, affiliates WXYZ.com, major league sports (MLB.com™) as well as an emerging class of Internet-centric producers such as Rocketboom™ and CNet™. Content on these sites is typically from a single source, but due to co-ownership of content, the user may observe several different "brands" such as television network call letters often owned by the same company. Content is usually posted to the Web after it has aired with the business intent to extract additional advertising revenue from the content. However, we are seeing a trend to simultaneous release of content on the Web and on traditional distribution mechanisms. A second goal is to generate more viewers for the next episode to be aired. Going forward, the affiliated local TV stations may become disintermediated and suffer revenue loss as viewers move from the channel based consumption model and get the content directly from the national "broadcaster." Network affiliates post local news and other content, and may receive content from the network for their site.

1.3.2 Aggregators

Sometimes called "Internet Broadcasters," aggregators act as centralized repositories that give users a wider range of content sources, with the goal of providing content providers with more viewers for their content. Samples include Brightcove™, ROO® Media and the FeedRoom™. These sites can be "white labeled" and branded by others such as broadband Internet service providers so users may not recognize these names.

A second, more widely recognized class of aggregators includes MSN™, AoL™, Google™ and iTunes™. Business models vary widely and include rental, purchase, advertising, and subscription. The primary access method is HTTP streaming, but some sites also offer higher quality video via managed download.

1.3.3 Download

Movie and video download sites such as MovieLink, Akimbo, and CinemaNow support search and allow users to rent movies. The media is good quality and is typically downloaded to local storage using a download manager which must be installed on the users' local machine. In addition to managing media file transfers to local storage, the client supports DRM which prevents copying the content and enforces the business rules for rental periods (e.g. keep for up to 30 days, play for 24 hours.)

1.3.4 Sharing

In addition to sites featuring professionally produced video, there are a large number of sites designed for sharing user generated content (UGC), or more precisely, "user contributed content" (UCC). The most widely known of these is YouTube™ which was purchased by Google™ for $1.6B in stock in 2006, however there are many others in use. In some cases these are not much more than network storage, but most add features such as transcoding to a common format, entry and search of author-supplied metadata, social tagging, and even e-commerce. (e.g. Putfile, Vsocial). The content on these sites is mainly consumer generated such as video blogs, but may include a range of qualities and genres. Unfortunately it may also include pirated copies of copyrighted material as well.

In addition to the aggregators, video download and video share sites, the traditional search engine model involving Web crawl for content discovery is evident. However, blind crawling has taken a back seat to feed based content pull using various syndication formats which describe the media at a high level via XML. Also, top search sites now cross-index to broaden coverage (e.g. Google™ video results will appear in MSN™ searches.) There are hundreds of Podcast search sites which offer none of their own content, but rather direct users to a broad range of content providers.

1.3.5 Application Specific

Vertical search sites that cater to a specific audience or to a narrow range of source content may include more structured metadata since the source of video is more controlled. For example, MLB.com™ offers video content from baseball games that have been highly annotated with detailed metadata indicating the player, the game situation, the stadium, etc. All of this data is accessible using an HTML forms interface to create very specific queries. Other sites such as IMDb or TVGuide® focus on video metadata such as movie information or guide listings, but may only contain preview clips rather than the full video content. In some cases, there is very little video available at all, and the sites may contain only related media such as photos of actors, box art, etc. The goal of these sites is to generate rentals or purchases in the case of movie sites, or to plan TV watching or schedule TV recording. On the other hand, the Internet Archive has a long history of offering both metadata and video for content in the public domain, or with very liberal copyrights.

1.3.6 Other Video Systems

There are an increasing number of IP applications for viewing Internet TV such as Joost™, or Miro™ which attempt to organize Web video feeds as channels whether they are in fact live streams or published as feed based media files. Video search is a key element for content selection, along with other means such as promotional placement or popularity rankings. Other video search applications may offer Web-based front ends, but require subscription. Media monitoring services allow subscribers to search and browse content as it was aired for various purposes including advertising verification, and corporate public relations. Media asset management (MAM) and digital asset management (DAM) systems are used for “in-house” production and archiving applications. These include workflow automation and may support Web-based distribution and monetization features, but at a minimum include a Web-based UI for administration, asset browsing and retrieval using metadata search.

1.4 Classification of Video Sources

As we discuss video search, it is important to keep in mind that the nature and quality of video varies widely depending on the application. The value of video can be difficult to judge; we can assess this on many dimensions such as image quality, or more subjective aspects such as educational, entertainment or historic value. Cost is more quantifiable, in fact, we can think of a "production cost spectrum" as shown in Fig. 1.1, where level of effort or cost of production vary from almost nothing to perhaps thousands of dollars per minute of final product for broadcast television content. Major motion picture costs can run even higher, particularly if we factor in the cost of promoting the project. Clearly, this huge range in content value has significant implications for Web search engine systems – it affects the content, quality, encoding, and availability of metadata and affects the degree to which automated methods can be employed to generate additional metadata to create index data.

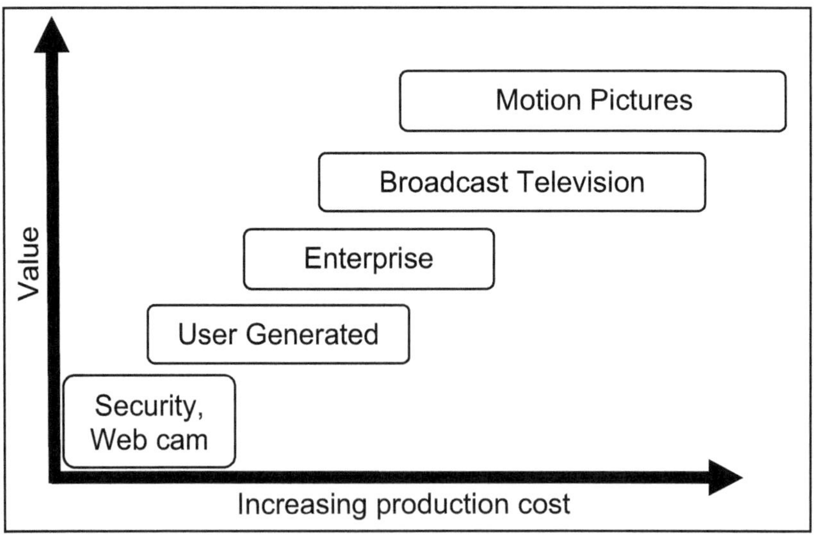

Fig. 1.1. Video production costs vary widely depending upon the application.

1.4.1 Webcams / Security

At the low end of the video production cost spectrum is content from automated cameras such as security or Webcams. These systems typically have some level of operator control, but the operator controls a large number of cameras, often from a remote location. As a result, they rely on automatic gain control (AGC) and are often oriented poorly with respect to available natural lighting. Optics and processing circuitry are low cost, resulting in poor imagine quality. Often the effective frame rate for these systems is reduced, perhaps even extremely so – to on the order of one image saved per minute. There is typically little or no camera motion, although some panning systems may be employed, and some views, such as from traffic cameras, may be affected by wind causing undesirable camera oscillations.

1.4.2 Video Telephony / Teleconferencing

Video telephony typically employs low cost terminal equipment, and relates to search when we consider video-mail systems. Semi-automatic systems are available for video conferencing that may include automated camera controls to follow the most active speaker, or remote and local camera control using a motorized pan, tilt, and zoom. Higher end conferencing systems feature high definition cameras and monitors, but image quality still suffers from poor room lighting conditions and lack of camera operators.

1.4.3 Industrial / Academic / Medical

Specialized systems for machine inspection for manufacturing quality control can run 24 hours a day, but much of the video is not stored. Video from ultrasound or other medical diagnostic equipment can be costly to produce due to equipment costs and skilled technicians or staff required. Remote sensing (satellite or high altitude reconnaissance) video may be very high resolution, and include telemetry data. Like other scientific applications such as microscopy, this video may be largely two dimensional, with little depth.

1.4.4 User Generated Content

Perhaps the lowest production cost for any manually created video is referred to as "user generated" content such as from mobile phone capture or from consumer-grade digital cameras. The cost of entry for these systems is miniscule – in some cases wireless service providers give users camera phones for free with a subscription. Of course, the cost per minute of video is related to the service charges, but for digital cameras, there is essentially no cost after the camera is purchased. While most users simply share or perhaps store videos in their personal media collections, some will go the next step and edit the clips into more palatable presentations. Free editing tools are available for both Windows and Mac platforms. The learning curve is very short and users can easily add transitions, titles, etc. Video editing is one of the most resource intensive applications, but most recently purchased PCs are up to the task. We are also seeing the emergence of online editing tools which remove the requirement of a powerful client PC since the editing is done on the server. Again, the cost per minute of produced video is low, but this assumes that the authors' time invested in editing the video is not valued. Video Blogs or Vlogs are typically amateur-produced content on a recurring basis and published to the Web, often with text commentary and these also fall into this category.

1.4.5 Public Access and Government (PEG) Content

In the US, local governments and community organizations such as high schools are given access to one or more cable channels to broadcast events for the benefit of the community and for educational purposes. Usually the production staff are not video professionals, and may even be students of broadcasting. The equipment may be better than consumer-grade, but it is semi-professional at best. This keeps costs low, but between this and the lack of experienced staff, it results in low quality output.

1.4.6 Enterprise Content

Corporations are increasingly using video as an additional means of employee communications as well as for training, and public relations purposes. The production may be outsourced or handled by a dedicated group. The content is produced using semi-pro or pro equipment (also known as

"Pro A/V") but often with a small staff serving multiple roles in the production process.

1.4.7 Rushes, Raw Footage

Professionally produced video relies on a formal workflow process, one stage of which involves creating several shots of each scene. The footage from this stage of the process serves as the raw material for the editing process. The quality is usually good, since professional grade cameras and good scene lighting techniques result in low noise images, and professional camera operators know to avoid the mistakes made by armatures (rapid unstable camera motion, automatic gain control artifacts, etc.) There is often a 10 to 1 ratio or higher of this content to the final product.

1.4.8 News

National news is expensive to produce and is typically of high technical quality, but due to demanding production schedules and live coverage, as well as the lower production budgets available for local news production, some artifacts may be present in the output. In fact, in some cases broadcasters may use low quality video from low bit rate links for feeds from extremely remote locations.

1.4.9 Advertising

Promotional video takes a wide range of forms, from the familiar 30 second spots all the way up through one hour infomercials to 24 hour shopping channels. In addition to marketing, public relations groups in corporations use video as an effective tool to get their message out. Archives and databases of advertising content are used for competitive analysis by corporations. While TV viewers loathe most ads, some have entertainment value, and the notion of targeted ads or telescoping where interested users interactively delve deeper into ads of interest may reduce the stigma of TV ads somewhat.

1.4.10 Episodic TV Programming

This category includes primetime entertainment programming, comedies, dramas, game shows, soap operas, etc. Within this category, there is a range of production costs with the assumption that better programming costs more, but generates more viewers and therefore more advertising revenue. Increasingly, we find that episodic content is made available via DVD for purchase (generally released such that only the previous seasons' episodes are available.) Content may be funded by subscription, publicly subsidized, or commercial – in which case the narrative flow will be interrupted with commercial breaks. For most commercial TV news and episodic programming, the entire program format and sequencing is driven by the placement and duration of commercial messages.

1.4.11 Feature Films

Again, there is a range here from independent ("indy") films or documentaries which may have a very low budget, all the way up to Hollywood movies such as Titanic which cost $200M or about $1M/per minute. In addition to major motion pictures for theatrical release, there is the second tier, with somewhat lower associated costs, such as made-for-TV movies and movies released on DVD-only. It is interesting to note that "digital cinema" is being developed for digital distribution and projection of movies, but the expensive installed base of film projectors as well as other factors has slowed its deployment.

1.4.12 Content Value

Within many of these types, there is a range of purposes: to inform, to entertain, and to persuade. The content reflects on its creators as well as its intended audience, their culture and value systems. Continually, creative thinkers strive to build novel experiences for audiences. Therefore these general classes should be understood to be only approximate, to give the reader a flavor of the range of video material encountered, and to provide an appreciation of the scope of the problem domain for video search engines.

Although we can estimate production costs for these content types, estimating the value for the user is more difficult. Security footage is largely valueless except for rare instances when a criminal act is captured, and then the value can have enormous impact. Home video content may be quite valuable for immediate family members, but of little value for anyone else. If a person gains celebrity, then video from their childhood may be of great interest for large audiences. News archives have incalculable historic value. In general terms, we may assess a searchable video collection on these merits:

1. production cost and percieved value of the contentl;
2. size and breadth of coverage of the content archive;
3. size of the audience interested in the content;
4. motivation for search (entertainment, research, forensic, etc.);
5. degree to which the content is accessible (on line, either open or restricted);
6. video quality (resolution, bit rate).

It is also important to consider the value of automated indexing systems, and here we draw a distinction between using media processing to derive information about the video contents and using manual methods to create this data. Even though it may be of great value to spot a terrorist in 10,000 hours of airport security camera footage, if there are no reliable algorithms to perform this search, then we cannot realize this potential value. Also, manually created metadata may be available to different degrees for each of these content types either via logging production data (e.g. the text of the titles typed into a consumer video editor) or by annotating post-production information such as with Major League Baseball statistics. For manually extracted data, a special purpose database is constructed, while a search engine must derive common tags from a wide range of content sources – and currently this metadata normalization is not a fully automated process. High value content benefits less from automated media logging or metadata extraction while for lower production budget content, these automated methods are more valuable since manual labeling is not practical. Since the value of the content falls off at lower production costs, the center of the production cost spectrum, semi-professional or enterprise content, represents an area of opportunity for video systems research [Chang02].

1.5 Challenges of Video Search

Searching requires browsing sets of candidate results. Video is a continuous (or linear) medium: if paused, only a single frame remains, audio is lost. Text is displayed in a more parallel fashion and can therefore be browsed easily. Video storage and transmission requirements are several orders of magnitude greater than those for text. Textual features (characters, words) are well defined, can be efficiently encoded, and are limited in number. Video features (edges, colors, motion) and acoustic features (pitch, energy) are less well-defined, computationally expensive to extract, and bulky to represent. In fact, there is little consensuses on which features are best for a given application. Furthermore, users can formulate textual queries easily using a keyboard so that, to a first approximation, the information retrieval problem reduces to a symbol look-up (i.e. find me the documents containing this word). For video databases, the query-response cycle is cross-modal (enter text, retrieve video). *Query by image content* involves building a query by specifying image or video attributes perhaps with a graphical tool which is beyond the patience limits of the typical user [Flick95]. *Query by example* or relevance feedback methods are easier to use but require some seed search to bootstrap the process.

Comparing some of the issues faced by video search engine systems to their analogs from the text domain sheds light on the nature and scope of the challenges encountered.

1.5.1 Acquisition

The term invisible Web or hidden Web refers to Web resources that are not easily indexed by Web search engines. Search engines use crawlers (also called spiders) to locate content for indexing by following links that they encounter in each page that they parse. However, instead of maintaining large collections of HTML files, many sites generate HTML pages dynamically from content stored in XML files or in relational databases. The content may be exposed only if users search using a Web form, an action which crawlers cannot easily mimic. Another problem for crawlers arises from sites that require user registration and authentication in order to access content. Estimating the size of the invisible Web is obviously difficult since by definition the content cannot be seen, but it may be orders of magnitude larger than the surface (visible) Web. There are also socioeconomic aspects to this issue since surface content is dominated by commercial enterprises and is funded largely by advertising, while hidden Web content is often premium, academic, etc. Some would go so far as to dismiss the invisible Web content entirely by saying that since users only use search engines to locate content then it does not matter if content exists out of the reach of their favorite search engine.

Although the scale is not easily quantifiable, as far as users' expectations are concerned, the phenomenon of invisible Web is more severe for video than for text. There are cases where Web pages contain links directly to static video files, but this is the exception rather than the norm. Video content is typically accessed through a player with complex scripting used to specify the video asset. Due to the size of the media objects and complexities of maintaining news content, asset management or publishing tools are typically used which are linked to databases. In some cases the publisher may have rights to publish the content only for a limited time. Professionally produced video entails high production costs and sites recover the investment through advertising or subscriptions. Video advertising via forced playlists also foils search engine crawlers. Video protected by digital rights management (DRM) precludes content based analysis. Attempts by search engines to circumvent any of these revenue-persevering schemes will not be received favorably by the content owners. Consumer produced content posted on sharing sites, on the other hand, is often open to all viewers for free and sites may have mechanisms to generate permanent links to videos. Crawlers may encounter these links on other sites and the links point back to a full page rather than directly to the video file. Stream saver or downloader tools have been developed to work around these issues.

Stale links arise from content being moved or deleted after a crawler has indexed the content. While this is a problem in both the text and video domains, it may be more likely for video files because large file sizes or rights issues may lead sites to remove content after a certain period of time.

1.5.2 Media File Formats

Considering parsing, one can build a very useful text search engine by dealing with only a single content source file format: HTML. As an afterthought, one could add support for Adobe PDF, Microsoft Word and perhaps one or two more, but these formats represent such a tiny fraction of the total available Web documents that users may not even notice their omission. The HTML format is designed to be easily parsable and although authors may create mal-formed HTML, there are many available error-tolerant parsers to choose from. Video, on the other hand, comes in a wide variety of formats and it is not clear which format is the most popular at any given time. New formats emerge, rise in popularity, and then may be knocked from the top spot as still newer formats gain popularity. Keeping up with these developments is a challenge for video search engines. Video container file format parsers and decoders are complex and often brittle so that relatively minor deviations from the video encoding standard may cause parsing failure. Decoders may be able to deal with only a subset of the permissible video encoding parameter space or only handle certain "profiles" (e.g. MPEG-4 simple profile) and may not be able to deal with others. Solutions have been built to address these issues but these solutions are complex to configure, administer and can be costly to operate at scale.

1.5.3 Data Transport

The data transport protocols for media are more diverse than for Web text. Again, crawlers need only implement the HTTP in order to cover most of the Internet content, with FTP being a distant second. In fact, there are many HTTP stacks implemented in many programming languages. HTTP streaming for video is gaining popularity, perhaps due to firewall issues, but video servers frequently use RTSP running over UDP to maximize throughput. UDP is a good choice for real-time video viewing, but the inherent possibly lost data packets will cause problems for automated indexing systems. Search engines for broadcast monitoring applications may need to grapple with ATSC or DVB access issues.

1.5.4 Browsing

When generating search results, search engines represent documents by metadata such as title and URL, but they also include a brief summary or extract to enable users to quickly determine if the document is relevant to their query. In the text domain, the operation of extracting representative text segments is straightforward. Regular expressions can be used to efficiently identify text segments matching the user's query terms, highlight them with markup, and to locate blanks between words to break up long sentences. More sophisticated processing can remove redundancy to form more meaningful extracts. In the video domain, extraction or summarization methods are not well defined and require complex video processing.

The time required to preview video limits the total number of search results that a user is willing to tolerate viewing. Evaluating relevance of a particular document is more time consuming with video than in the text case. Neglecting HTTP site response time, text documents load within a second or two and users may be able to judge instantly if the page is worth reading, and if so, quickly spot-checking several points in the document is usually enough to determine if the document satisfies the query. For video, a much larger amount of data must be downloaded and buffered prior to start-up. After the video starts, the relevant content that the user is looking for is typically not in the first few seconds of playback. Video is normally consumed in a lean-back mode and so the content creators devote more time to lead-in material to pique the viewer's interest. If a viewer attempts to seek past this content, then re-buffering must take place, and it is unlikely that the desired location will be arrived at on the first attempt. The long lead time required to evaluate document relevance frustrates users of video search.

1.5.5 Duplication

Duplicate or near duplicate pages in Web search results can frustrate users as they repeatedly see pages that they have already rejected as being irrelevant to their query intent. In the text domain, duplication is trivial to detect and there are well accepted methods for determining document similarity (e.g. based on edit distance) that are reasonably efficient to compute in order to detect near duplicates. Duplicate videos in query results lists present even more of a problem for video search engine users. Videos take a significant amount of time to start playing and the delay will be intolerable for users if they encounter duplicates in query result sets. Sometimes cues from metadata and thumbnails will be enough for users to determine duplications, but not always. Duplicates are common in the video search applications, since a single source of video, say a television broadcast, may be captured by several viewers and posted to numerous sites. Also, the same video may be broadcast repeatedly or at different times for different television markets, so even if the recording time and broadcast channel of a captured video clip is available and accurate, that may not be enough to determine if the content is duplicated. Twenty four hour news channels often rebroadcast footage of breaking news and may intersperse this with new video as it becomes available. Video duplicate detection is an algorithmic challenge and proposed algorithms are computationally intensive. Often a duplicate clip is posted to sharing sites with differing metadata.

1.5.6 Ranking and Indexing

Text information retrieval including ranking and document indexing algorithms are mature, and off-the-self solutions that perform efficiently at scale are available. Video indexing is an emerging technology and universally or widely accepted techniques are not available and may not operate with the scale necessary for practical Web video search. Often the algorithms are domain-specific and cannot be applied to unknown arbitrary video content.

1.6 Advantages of Video Search over Text

Given all these aspects which make video search more difficult than text search, together with the fact that text search engines are far from perfect themselves, it may be surprising that successful video search systems have been deployed at all. This may be explained by considering areas where video search is less problematic than text search.

Although browsing video results sets is more time consuming than for text, the human visual system can process images more quickly than text. Therefore a first level of results set filtering can be nearly instantaneous. Of course this assumes that a reasonable set of representative key frames have been extracted and can be rendered quickly. Users can scan arrays of these images to quickly select potentially relevant video segments. Obviously this process is not without error since a single key frame cannot convey all of the information from the video clip which is of course a sequence of frames. However, users can make reasonably accurate general assessments of the global nature of the video given a single frame. For example, one can differentiate easily between broadcast television content and amateur video blog postings based on the quality and content of a single frame, particularly if certain cues such as text overlays are present.

1.6.1 Applications

Another factor in video search engines' favor relates to the application areas and user expectations. Often video search is used for entertainment purposes in which an irrelevant video may be less of a problem than in the text domain. Text search can also be used in a less task-oriented, more entertainment-like mode where the user meanders in different directions than the original search topic. With video search, however, the user fully expects that consuming the results of the search will take time given the linear nature of the media. In this sense video search is a more forgiving task than text search and users may be more tolerant of error in some applications. On the other hand, applications including education or research are not error-tolerant and even entertainment applications will be improved given more accurate or personalized video search and some controlled semi-randomness in the results set can be injected if desired. Search activities can be classified into three broad categories: (1) browsing or exploring the collection; (2) finding an arbitrary video that satisfies the query; and (3) finding all relevant videos [Camp07].

1.6.2 Metadata

One further area where video search engines may have a slight edge over text search engines concerns metadata. Good quality video takes more effort and budget to produce than text documents and therefore video producers may be more likely to take the time to include metadata descriptions such as genre classifications, plot synopses, and keywords. In this sense, video search is more like book search. Note that this is not the case for video blog content and that as video acquisition and editing technology improves, it will be even easier for more, less diligent, users to create video content. It is also true that documents such as books and journal articles are time consuming to produce and similarly warrant the inclusion of abstracts, keywords and subject matter classification.

1.7 Metadata vs. Content

Metadata is "data about data," or in this case "data about media." *Global metadata* refers to the entire media assets and typically includes a title, author, copyrights, etc. While almost every video application supports global metadata in some form [DC03], for some applications additional sets of metadata may pertain to segments of the media – such as for news content were one segment may contain footage from a third party and copyrights may be more restrictive for that segment.

1.7.1 Content-based retrieval

Content-based retrieval involves the use of metadata that is derived (typically automatically) from the media streams and almost always includes a temporal attribute. Note that a transcript of the dialog is usually considered to be "content" and not metadata although it can be represented concisely in data structures similar to those that represent global metadata such as the description, and a text stream can be used for both consumption and navigation.

Most video searching systems rely on high level attributes accompanying the video files for search. These typically include title, date, genre, and brief description. However the next generation of searching systems goes beyond metadata to search the content of the video. Content indexing not only provides a more accurate description of the media; it supports temporal information so that users can navigate to the desired segments of long-form video material. It is content indexing that enabled Web search engines to excel since codifying and maintaining consistent document metadata is impractical, from both the engineering and social perspectives. Video search systems must leverage existing global metadata, incorporate any manually added detailed metadata or tags, extract additional detailed metadata automatically, and support the on-going addition of viewer supplied metadata such as tags, ratings, comments, and popularity.

1.8 Conclusion

The amount of video content on the Web is growing rapidly as new technologies such as Internet protocol television (IPTV) and mobile video are deployed. Video search engines are being developed to enable users to take advantage of these video resources for a wide variety of applications including entertainment, education and communications. However, the task of information extraction from video for retrieval applications is challenging, providing opportunities for innovation.

All video is not created equal; there is a wide range in terms of quality, available metadata and content. We described some of the challenges for video search as related to text search, and introduced the notion that metadata plays a key role in the accuracy and effectiveness of video search. The metadata may accompany that content and be easily ingested for search, and powerful media analysis technologies may be employed to extract additional, detailed metadata for search. Users may be participants in the metadata creation process through tagging and otherwise commenting on the video that they have viewed. Analysis of user activity can lead search engines to make implications about video content and quality or popularity.

References

[Camp07] Campbell, M. et al.: IBM Research TRECVID-2007 Video Retrieval System, *TREC Video Retrieval Evaluation Online Proceeding,* National Institute of Standards and Technology (2007).

[Chang02] Shih-Fu Chang, The Holy Grail of Content-Based Media Analysis, *IEEE Multimedia Magazine*, **9**(2), pp. 6–10 (2002).

[DC03] Information and Documentation – The Dublin Core metadata element set, ISO Draft International Standard 15836:2003 (2003).

[Flick95] M. Flickner, et al.: Query by image and video content: the QBIC system, *Computer*, **28**(9), pp. 23–32, Sep. 1995.

[Haupt05] A. Hauptman, Lessons for the Future from a Decade of Informedia Video Analysis Research, *Lecture Notes in Computer Science*, Springer, Berlin / Heidelberg, vol. 3568, pp. 1–10 (2005).

2 Video Data Sources and Applications

2.1 Introduction

To further illustrate the challenges and opportunities for video search, this chapter will address the nature, availability, and attributes of different sources of video data. Search engines leverage all available information relevant to media and this chapter will provide details about available metadata for different types of video including electronic program guides, content identifiers, video on demand packages, and syndication standards. We will also introduce representations of textual information associated with media such as transcripts, closed captions, and subtitles. The breadth of metadata sources is described at a high level, and more detailed information is provided for selected domains including, broadcast television, digital video recorders (DVRs), consumer video, and Internet sources such as podcasts and video blogs. After media is published on the Web, additional metadata may accrue from social sources in the form of tags, ratings, or even user contributed subtitles, all of which can be exploited by video search engines to produce more accurate results.

2.1.1 Evolution of Digital Media Metadata

From planning, through production, editing, distribution, and archiving, metadata is used throughout the video life cycle to manage, locate, and track rights and monetize video content. Historically, in the days of film and analog video, the tools used for this process were primitive, consisting of handwritten notes with labels and numbering schemes for tapes. As video has gone digital, so has the metadata as well as the production and distribution processes. This migration is not totally complete: film is still the dominant archival format for some content classes and there are legacy archives in a range of tape formats both analog and digital.

2.1.2 Consumer Video Metadata

On the consumer video side, the equipment costs and time scales are radically different. Even here, while photography has gone digital, and most cameras can capture video, digital camcorders have only recently begun to replace analog models. This move to digital consumer media capture has resulted in some limited benefits related to metadata. Looking at the file times of their personal media archives, consumers can tell the exact time and date of photos and videos that they've shot. However, the filename is typically only an obscure sequence number and the content of the media is only revealed when it is played. New devices, including camera phones, record each shot in a separate file, which is great for determining the time of the shot, but only adds to the problem of locating particular content due to the sheer volume of files created. In this respect, consumer video metadata is not much farther along than the pencil and paper days of video production. Some promising developments on this front include video cameras that log start and stop to provide a shot index, support GPS (Global Positioning System) information as well as advances in consumer video editing packages that encourage the addition of titles and other metadata.

2.1.3 Metadata Loss

Metadata is of critical importance for search systems and it is important to capture all available metadata because in addition to the objective media parameters, good metadata tagging captures the subjective essence of the media. It is often how users refer to content (by title, actors, etc.) Today's automated content analyses tools can augment this data, but cannot extract the high-level semantics reliably. Unfortunately, in many cases metadata is lost somewhere in the process from capture to delivery to end users. Video asset management systems preserve and manage metadata as well as the media throughout the content life cycle. The typical asset life cycle includes not only preproduction, editing / post-production, and publication or distribution but also archiving and reuse for future production cycles (see Fig. 2.1.) The problems for metadata integrity and preservation arise as media flows from one system to another in this process. The production systems typically are assembled over time from various vendors and may involve handoff of content between several organizations, each with their own policies and practices for metadata.

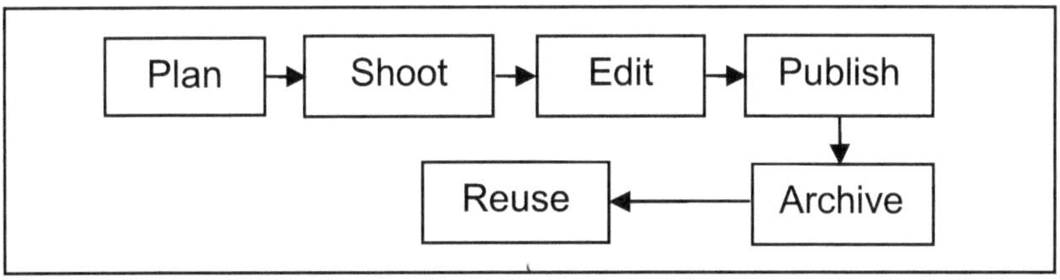

Fig. 2.1. Stages in the video content lifecycle. Metadata can be created and captured at each stage, but may not be preserved throughout the process.

Convergence of the TV and the PC is underway, and has been for 10 years or more, but the broadcast and Internet video communities still represent two distinct camps, and this has implications in the area of metadata standards. Having been designed for a medium that is international and online by nature, the Internet standards are most accessible for search engines and are based on technologies such as XML that are familiar to Internet engineers and application developers. Today, the concept of converged services in the telecommunications industry refers not only to TV and PC but also includes mobile handsets and connotes seamless access to media services across multiple devices and network connection scenarios. IPTV and the move to file based video production workflows will help bring the IP and broadcast communities closer together.

2.1.4 Metadata Standards

Metadata standards have emerged to facilitate exchange of media and its description between organizations and among systems and components in the video production, distribution, and archiving processes. Unfortunately, different industries, communities and geographic regions have developed their own standards designed and optimized for their own purposes. Therefore a truly universal video search engine must deal with a wide range of source metadata formats. Table 2.1 lists some of the metadata standards and the responsible bodies that have a bearing on video search applications, but there are many others such as ATSC (e.g. PSIP), ETSI, DVB, and ATIS/IIF. A detailed treatment of broadcast metadata is beyond our scope but we will introduce some of the metadata systems and interested readers can consult the references for more information (e.g. [Lug04]),

Table 2.1. Representative metadata standards.

Standard	Body
MPEG-7, MPEG-21	ISO / IEC – International Standards Organization / International Electrotechnical Commission, Motion Picture Experts Group
UPnP	Universal Plug and Play Forum (MPEG-21 DIDL-Lite, etc.)
MXF, MDD	SMPTE – Society of Motion Picture and Television Engineers
AAF	AWMA – Advanced Media Workflow Association
ADI, OCAP	CableLabs® (includes VoD)
TV-Anytime	Industry Forum, now ETSI – European Telecommunication Standards Institute
Timed Text	W3C, 3GPP, (MPEG)
P/Meta, BWF	EBU – European Broadcasting Union
Dublin Core	DCMI – Dublin Core Metadata Initiative, US NISO Z39.85, ISO 15836, OAI
RSS	Harvard
Podcast	Apple
MediaRSS	Yahoo
ID3	Informal

2.1.5 Dublin Core

The Dublin Core Metadata Initiative defines a set of 15 elements used for a wide range of bibliographic applications [DC03] and many media metadata systems incorporate the element tag names or a subset of the core elements. (The name refers to the origin of the initiative at the Online Computer Library Center workshop in Dublin, Ohio in 1995.) The core elements are easily understood, and are widely used in media metadata applications such as RSS Podcasts and UPnP item descriptions through the incorporation of XML name space extensions. Also, the Open Archives Initiative which promotes interoperability among XML repositories utilizes the DC element set [Lag02].

The core elements are not sufficient for most applications and can be thought of as a sort of least-common-denominator for metadata. For example, the Dublin Core defines a "date" element, but media applications may need to store multiple dates: such as the date that original work was published, the date performed, the date broadcast, etc. The Dublin Core element set has been extended (and referred to as the Qualified Dublin Core) to address this [Kur06]. There are three XML namespaces that define DC encoding (Table 2.2).

Table 2.2. Dublin Core metadata and namespaces.

Elements	Namespace
The 15 Dublin Core Metadata Elements (DCMES)	http://purl.org/dc/elements/1.1/
DCMI elements and qualifiers	http://purl.org/dc/terms/
DCMI Type Vocabulary	http://purl.org/dc/dcmitype/

2.1.6 MPEG-7

MPEG-7 is an ISO/IEC specification titled the "Multimedia Content Description Interface" with a broad scope of standardizing interchange and representation of media metadata from low-level media descriptors, and up through semantic structure [ISO/IEC 15938]. Further, content management, navigation (e.g. summaries, decompositions) as well as usage information and user preferences are covered. Unlike some other metadata standards, MPEG-7 is not industry specific, and it is highly flexible. There are over 450 defined metadata types, and many type values can in turn be represented by classification schemes [Smi06]. To promote a wide range of applications and to allow for media analysis algorithm development, MPEG-7 does not specify how to extract or utilize media descriptors, but rather it focuses on how to represent this information. For example, visual descriptors include contour-based shape descriptors for representing image regions, but no assumptions are made about preferred image segmentation algorithms. MPEG-7 components are used in other metadata systems such at TV-Anytime and MPEG-21.

2.1.7 MPEG-21

MPEG-21 defines packages of multimedia assets and promotes interoperability of systems throughout the asset value-chain [ISO/IEC 210000.] The concept of a digital item (DI) is introduced as well as a digital item declaration language (DIDL). MPEG-21 Digital Item Identifiers (DII) serve the purpose of uniquely identifying content, and incorporate the use of application-specific identifiers (such as ISRC, see Sect. 2.3.6) rather than specifying yet another competing system. MPEG-21 is broad in scope addressing a broad range of practical issues encountered in monetizing media assets such as DRM, and media adaptation for various consumption contexts.

The concept articulated in the MPEG-21 Digital Item Adaptation (DIA) is particularly interesting for video search engine systems and services that are built around them. The ultimate goal is that of promoting "Universal Multimedia Access" (UMA) to allow content producers to create a unified media item or package and allow viewers to receive the content on any device at any time. While fully automating this process may not yield ideal results, the standard allows content creators to specify in as much detail as possible, the manner in which the content is to be adapted. Not only are the resources adaptable as would be expected, but also the descriptions of those resources are adaptable as well. Adaptation is possible at both the signal level (e.g. media transcoding) as well as at the semantic level. One aspect of adaptation involves the terminal capabilities such as available codecs, input/output capabilities, bandwidth, power, CPU, storage and DRM systems supported by the device. Beyond this, MPEG-21 supports adaptability for channel conditions, accessibility (for users with disabilities), as well as consumption context parameters such as location, time, and the visual and audio environment. Of particular interest for the current subject of media metadata that we are considering is the MPEG-21 notion of "metadata adaptability." Three major classifications are brought to light:

1. Filtering – a particular application will use only a subset of all of the possible metadata available for a particular digital item;
2. Scaling – reducing the size or volume of metadata as required by the consumption context (e.g. bandwidth or memory constraints);
3. Integration – merging descriptions from various sources for the digital item of interest.

MPEG-21 supports many more capabilites relevant for video services, such as session mobility. Interested readers are referred to [Burnett06] from which we have drawn to provide a breif introduction to this topic.

The Universal Plug and Play (UPnP) specification defines a "DIDL-Lite" which is a subset of the MPEG-21 DIDL [UPnP02]. This is an example of the incorporation of multiple metadata specifications since the Dublin Core elements are also supported via XML namespace declarations in addition to the defined "UPnP" namespace which augments the DIDL-Lite to form the basis of the specification. Further, this mechanism can be extended, for example, to include "vendor metadata" such as DIG35, XrML or even MPEG-7 [UPnP02].

2.2 Essential Media Metadata

2.2.1 Embed Global Metadata

Some level of metadata is embedded within the media stream, either as a header for use in decoding or rendering, or as an additional logical bit-stream multiplexed within the media. In addition, metadata can be maintained in separate files that refer to the media file or collections of files (packages or channels) see Fig. 2.2.

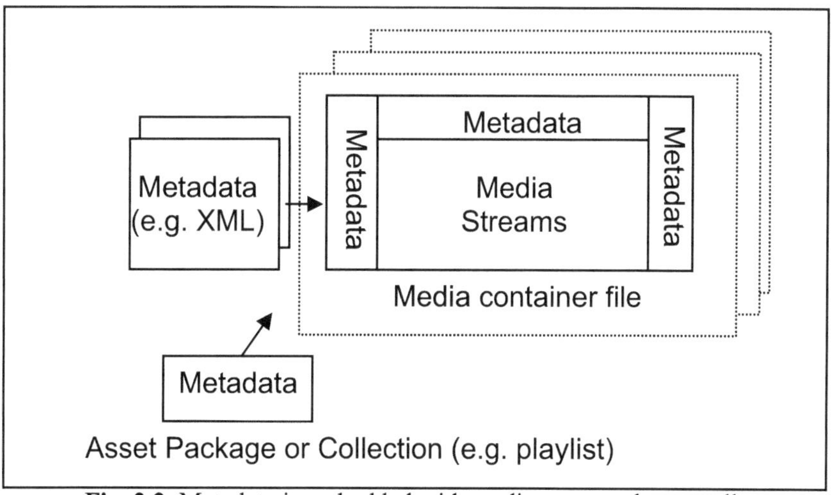

Fig. 2.2. Metadata is embedded with media, or stored externally.

2.2.2 Elementary Metadata

There are several common media container formats in use on the Web today, and search engine ingest systems can parse these to extract some basic information about the media. At a minimum, the container files indicate enough global information about the media to allow decoding applications to play the file (e.g. the number of media streams in the file.) Each stream has attributes such as bitrate, frame rate, spatial resolution, etc. Typically there is at a minimum an audio stream and a video stream, but there may be multiple media streams, e.g. multirate streaming uses several video streams at different bitrates. The duration and image resolution information are required for parsing and can be of some limited practical use for search engine systems for filtering search results (e.g. "find me videos with at least a 320 by 240 resolution and 20 minutes or more in length").

The container formats typically support the inclusion of high level descriptive information such as title, publisher, etc. This global media metadata can be extracted from open formats using available tools and used for indexing applications. This provides some very basic information for search engines beyond the typical filesystem attributes such as size and name, but the level of detail available falls short of providing a true description of the content itself. We can think of levels of depth of information discovery about unknown media files as shown in Fig. 2.3.

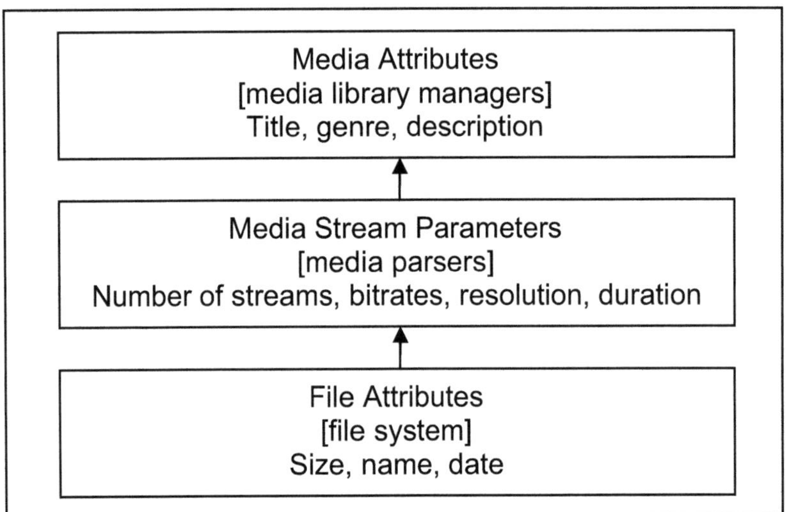

Fig. 2.3. Depth of discovery of metadata from media files.

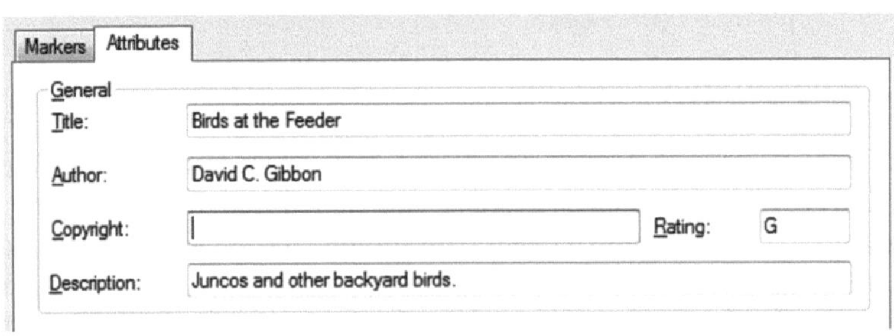

Fig. 2.4. Global metadata entry dialog for a single media file (Microsoft® Windows Media® Editor).

2.3 Metadata for Personal Media Collections

2.3.1 Consumer Media Libraries

Fig. 2.4 shows a typical user interface for authoring this level of metadata. In this example, these represent fields in the Content Description Object in Microsoft's ASF specification [ASF04]. While these are five core attributes used by many Microsoft applications [Loomis04], ASF also supports an Extended Content Descriptor Object to include other name–value pairs and there is also a Metadata Library object for including more detailed information. Most formats support arbitrary name–value pairs but applications promote or require the use of standard attribute taxonomies. Also, as a practical matter, typical users are not aware of metadata unless the values are displayed by the applications that they use on a regular basis. Media players may make this visible through a "media properties" or "media info" dialog, but often this is not readily accessible from the top level user interface of the player application. However, with the emergence of personal media library manager applications, users become very aware of the metadata and may even be motivated to edit and maintain the data for their collections, but will certainly favor systems that manage this for them.

Fig. 2.5. Metadata for personal media collection (Apple® iTunes®) including items with missing metadata.

A sample media library metadata view is shown in Fig. 2.5, in which users locate content by sorting and browsing or search. A subset of the metadata fields is shown and the interface allows additional fields to be chosen (overlaid menu in the figure.) However many of these fields are of little value for search (e.g. beats per minute) or are only suitable for specific classes of content (e.g. episode ID). As can be seen in the figure, not all fields are fully populated, and there are often inconsistencies in this type of data.

Of course, all media library managers do not use the exact same set of metadata tags so if a media search engine were to ingest content using tags maintained in personal media libraries, it must perform translation or mapping of metadata tags. For example, Table 2.3 lists two major consumer media library applications tag names, and suggests a mapping. Note that metadata mapping, while trivial here, is typically much more difficult and in most cases shades of meaning are lost in this normalization process.

Table 2.3. Media library mangers label attributes differently (only a few representative differing tags names are shown; most tag names are the same).

Windows Media Player 10	Apple iTunes 7
Length	Time
Type	Kind
Mood	N/A
N/A	Equalizer

Libraries are maintained in persistent databases which may expose interfaces to allow other applications to interoperate. Alternatively, import and export of library data can be used for this purpose. For example, Apple's iTunes exposes library data in a straightforward key–value XML format as shown in Fig. 2.6.

```
<key>Kind</key> <string>MPEG-4 video file</string>
<key>Size</key>         <integer>40761183</integer>
<key>Total Time</key>   <integer>524864</integer>
<key>Year</key>         <integer>2007</integer>
<key>Date Modified</key>
   <date>2007-05-17T20:48:13Z</date>
<key>Date Added</key>
   <date>2007-05-17T20:48:06Z</date>
<key>Bit Rate</key>       <integer>127</integer>
<key>Sample Rate</key>    <integer>48000</integer>
<key>Release Date</key>
   <date>2007-05-09T22:00:39Z</date>
<key>Artwork Count</key>  <integer>1</integer>
```

Fig. 2.6. Media library metadata for a single asset (extract from Apple® iTunes® library XML format).

2.3.2 UPnP Forum

Media library managers typically monitor the local filesystem, or a set of specified folders for the addition of new media and maintain a database of extracted metadata from media file formats that they support. While it is possible for consumers to monitor collections of media on other computers in their home network, configuring file sharing can be cumbersome due to system incompatibilities and security configurations such as firewalls. The emergence of easy to use, low cost, high capacity network-attached storage devices promotes the concept of shared media storage. To enable ease of use and to foster interoperability of networked media devices, the computing, consumer electronics and home automation industries formed the Digital Living Network Alliance (DLNA) and the Universal Plug and Play (UPnP) Forum which defines a range of standards and defines the concepts of Media Servers and Media Renderers [UPnP02.] UPnP uses the Digital Item Declaration Language defined in MPEG-21 [DIDL01] and in particular defines a subset referred to as DIDL-Lite.

2.3.3 MP3 ID3

ID3 tags arose out of a need for organizing MP3 music files into libraries for so called "jukebox" applications. Note that ID3 is not part of the MPEG specifications, but rather it is a means of appending data to MP3 files. ID3v2 supports not only global metadata, but also detailed metadata and even supports embedding of images. Beyond metadata related to the asset, ID3 supports application specific features such as encoding the number of times that a song has been played. As flexible as the ID3v2 format is, the precursor ID3v1 was extremely limited and rigid. (However, this utilitarian simplicity resulted in wide adoption.) The fields used in the ID3v1 Tag are shown in Table 2.4, and include song title, artist, album, year, comment, and genre. The strings can be a maximum of only 30 characters long and the genre is an 8 bit code referencing a static table of 80 values. Later this was extended to 148, and in ID3v1.1, the comment field was shortened to 28 characters and a track number was added.

Table 2.4. A subset of the Dublin Core metadata elements and their analogs in popular media file formats.

Dublin Core Elements	Quick-Time	Mobile MP4	MP3 ID3v1	Microsoft ASF base
Title	nam	titl	Song Title	Title
Creator	aut,art	auth	Artist	Author
	alb		Album	
Date	day		Year	
Description	des,cmt	dscp	Comment	Description
Type			Genre	
Rights	cpy	cprt		Copyright Rating

2.3.4 3GP / QuickTime / MP4

The QuickTime file format [QT03] was adopted for the MPEG-4 file format (MPEG-4 part 14) and is sometimes referred to by its file extension MP4. Quicktime also uses the extension MOV for a wide range of media formats. The 3rd Generation Partnership Project (3GPP) defined the 3GPP file format (3GP) which is essentially a specific instance of MP4 with some extensions for mobile applications, such as including support for Adaptive Multi-Rate (AMR) format audio. It is used for exchanging messages using the MMS protocols. Metadata tags including author, title and description are defined in asset metadata within a user data "box" (ISO file format structure segment) [3GPP, p.29]. In addition to these fields, a box is defined to store ID3v2 tags directly without translation or mapping required.

2.3.5 Metadata Services

In addition to parsing media files to extract metadata, personal library managers gather additional metadata from online repositories. The basic data flow is shown in Fig. 2.7. An application on the client reads an identifier from the local media or, in this example, calculates a unique identifier based on the length and number of the tracks on a compact disk, and then requests additional metadata from a server using this identifier as a query. The response returned can include detailed up to date metadata and images of cover art (or box art.) Library applications can also calculate signatures given a single media file, such as an MP3 file. The signature is used to query a database of signatures to determine a unique content identifier. This form of content identification is sometimes called "fingerprinting" and it has been used successfully to detect copyright infringement on music sharing services. User-generated content video hosting sites such as YouTube are notorious for hosting copyrighted content posted illegally by users. The sites take refuge from liability in the "safe harbor" provisions of the Digital Millennium Copyright Act [DMCA98], but are increasingly using fingerprinting to identify and remove copyrighted content. Fingerprints based on the audio component are useful for A/V applications as well – particularly in cases where users dub copyrighted audio into a video that they are producing. However, video fingerprinting systems are more recently available.

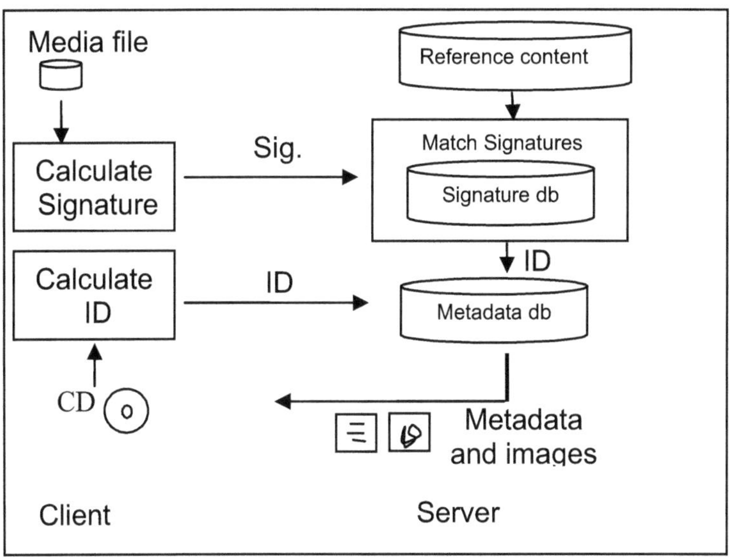

Fig. 2.7. Metadata services based on content identification.

Providers of this class of media metadata services include Gracenote® (formerly CDDB), Freedb, and All Media Guide AMG (allmediaguide.com). For movies, the Internet Movie Database (IMDb) maintains a detailed collection of metadata for over 500,000 items (movies, TV episodes) and includes plot summaries, actors, and directors, in addition to typical metadata such as genre, rating, and title. JPEG images of "box art" are also available. The database is available for download for non-commercial applications, and imdb.com provides some basic tools for querying the database locally. Other entities such as AMG maintain similar movie metadata databases and provide subscription based services for access. Many of these services also maintain Web sites for end users to search and browse media metadata, some with a business model of driving sales of hard media (DVDs or CDs.)

2.3.6 Content Identification

Content identification is a critical component for these systems and several unique content ID (CID) methods have been developed for this purpose. Fingerprinting is still required for cases where the ID is not trusted, such as in user contributed video hosting applications, or for content systems that do not make use of IDs such as MP3 file sharing. Also fingerprinting systems use content IDs to link the signatures back to the source content metadata. Content identification is a general function and many standards incorporate some form of CID. In ISO, the TC 346/SC 9 group develops documentation identification standards such as ISBN, ISAN, and several others.

- **ISAN:** The International Standard Audiovisual Number (ISO 15706) is a system for uniquely identifying an asset, independent of the broadcast schedule or recording medium. MPEG-2 and MPEG-4 have a field for ISAN and the SMPTE Metadata dictionary supports ISAN as an identifier/locator. The ISAN has a standardized length as does the ISBN.
- **UMID:** Unlike the ISAN which is intended for the entire work as a unit, UMID Unique Material Identifiers (specified in SMPTE 300M) are used in Material Exchange Format (MXF) and can reference shots and even individual frames. Rather than being issued by an organization, UMIDs can easily be generated by camcorders in a manner similar to the way SMPTE timecodes are generated. Extended UMIDs go beyond identification and encode metadata such as the creation date and time, location and the organization.
- **ISRC:** The International Standard Recording Code, ISO 3901, is used for sound and music video recordings and includes a representation of country of origin, recording entity (record label), year and a 5 digit serial number.
- **CID:** This content identifier originated in Japan and is a product of the Content ID Forum or cIDf. The CID is designed to be embedded within the digital object, typically video or images.
- **CRID:** TV-Anytime defines the Content Reference Identifier or CRID to uniquely identify content independent of its location or instance. It is used for EPG applications. While typically refering to a single program, CRIDs may also refer to groups of programs or segments of programs.

- **ISWC:** The International Standard Musical Work Code, ISO 1507, is intended to uniquely identify a musical work for rights purposes as distinct from the particular recording for which the ISRC is intended.
- **ISMN:** The International Standard Music Number (ISO 10957) can be veiwed as a subset of the ISBN for works related to music such as scores and lyrics.
- **DOI:** Digital Object Idendifier (IDF – International DOI Foundation, the syntax is defined in ANSI/NISO Z39.84) offers a system for persistent identification and management of digital items. The strings are opaque unlike CID, ISRC, etc. Initally designed for text documents.
- **GUID:** Some systems such as RSS support a GUID or Globally Unique Identifier element. In RSS it is an optional string of arbitrary length, and some publishers use the URI of the media that is being published for this purpose. RSS will be explained in detail later.

2.3.7 Recorded Television

Personal media library managers used on media center PCs such as the Windows Media Center® and MythTV (Linux) organize recorded television by leveraging the EPG metadata pulled from service providers as shown in Fig. 2.8. The names of the columns across the top including series, episode, length (run time), description, genre, and rating are representative, and many other fields are available to help users find and manage their personal content collections. As is typically the case with metadata sources, some of the data is absent (e.g. episode description) or questionably formatted (e.g. the rating value of "**;PG;TV-14" seems to be a composite of multiple rating systems: MPAA, ATSC, and popularity rating).

Standalone DVRs, or DVRs integrated into set-top boxes may include a networking capability to support media sharing in the home network. The library manager can display the recorded program metadata, however rights issues may restrict this sharing to only other DVRs in the home (a service feature called "whole home DVR"). Fig. 2.9 represents the dataflow for both the EPG metadata and content in a typical DVR application. Although not shown in the figure, for recorded movies, a combination of the EPG metadata and data from Internet providers such as IMdb or AMG® as discussed above may be used. We will focus on EPG in detail later.

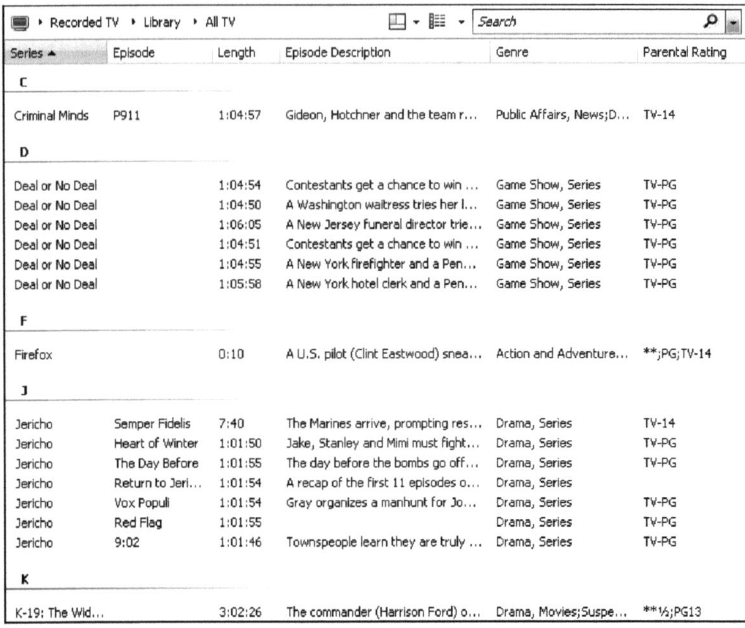

Recorded TV ▸ Library ▸ All TV — Search

Series	Episode	Length	Episode Description	Genre	Parental Rating
C					
Criminal Minds	P911	1:04:57	Gideon, Hotchner and the team r...	Public Affairs, News;D...	TV-14
D					
Deal or No Deal		1:04:54	Contestants get a chance to win ...	Game Show, Series	TV-PG
Deal or No Deal		1:04:50	A Washington waitress tries her l...	Game Show, Series	TV-PG
Deal or No Deal		1:06:05	A New Jersey funeral director trie...	Game Show, Series	TV-PG
Deal or No Deal		1:04:51	Contestants get a chance to win ...	Game Show, Series	TV-PG
Deal or No Deal		1:04:55	A New York firefighter and a Pen...	Game Show, Series	TV-PG
Deal or No Deal		1:05:58	A New York hotel clerk and a Pen...	Game Show, Series	TV-PG
F					
Firefox		0:10	A U.S. pilot (Clint Eastwood) snea...	Action and Adventure...	**;PG;TV-14
J					
Jericho	Semper Fidelis	7:40	The Marines arrive, prompting res...	Drama, Series	TV-14
Jericho	Heart of Winter	1:01:50	Jake, Stanley and Mimi must fight...	Drama, Series	TV-PG
Jericho	The Day Before	1:01:55	The day before the bombs go off...	Drama, Series	TV-PG
Jericho	Return to Jeri...	1:01:54	A recap of the first 11 episodes o...	Drama, Series	
Jericho	Vox Populi	1:01:54	Gray organizes a manhunt for Jo...	Drama, Series	TV-PG
Jericho	Red Flag	1:01:55		Drama, Series	TV-PG
Jericho	9:02	1:01:46	Townspeople learn they are truly ...	Drama, Series	TV-PG
K					
K-19: The Wid...		3:02:26	The commander (Harrison Ford) o...	Drama, Movies;Suspe...	**½;PG13

Fig. 2.8. Media library metadata (Windows® Media Player® 11 on a Media Center PC).

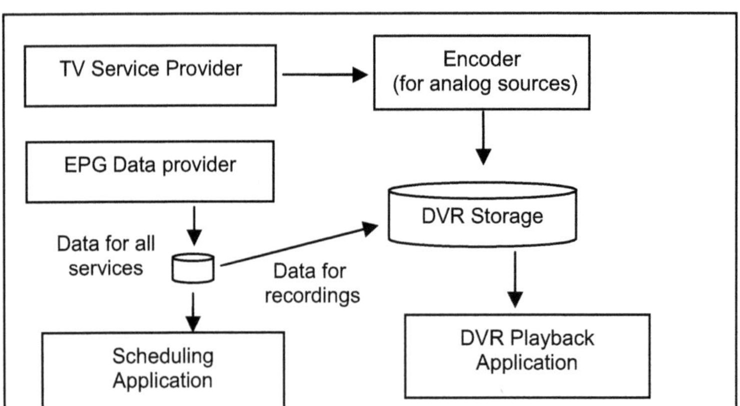

Fig. 2.9. Data flow for DVR EPG metadata.

2.4 Media Syndication: RSS Content Description

2.4.1 Content Syndication

RSS (really simple syndication) descriptors were developed for syndication of HTML news feeds but have evolved into a de facto metadata standard for Web media. RSS allows content producers to express high-level metadata about their content in a standardized way to enable other sites, aggregators, to display content from many different providers through a uniform user interface. Since Web revenue models depend on ad impressions, it may seem counterintuitive that a content provider would wish to syndicate their content, which would allow the aggregators to become Web destinations and therefore eat into the ad revenue stream. However, the primary purpose of RSS is to announce content and to entice users to visit the origin site. The content descriptors always include a URL (rendered as a clickable link by aggregators) to allow users to retrieve the content. Importantly, the content itself is not displayed on the aggregator's site where the aggregator would benefit from ad impressions, perhaps in a peripheral frame around the content display. Rather, the user is directed back to the content origin site, and this site is designed to promote other content from the same site, therefore increasing dwell time and ad impressions. Therefore, it can be seen that it is in the best interest of content providers to publish these RSS format content descriptions and to promote these to aggregators to maximize viewership. RSS is designed for recurring content or series and is organized as "channels" of "items." Note that the RSS "channel" is similar to a television series or program, not a television "channel." RSS "items" are similar to television "episodes." *The New York Times* does not have it's own single RSS representation, but rather has several RSS feeds, such as "technology news," "sports," "world news," etc.

2.4.2 Media Enclosures

Another important aspect of RSS is that in addition to the dynamic online mode of consumption described above, RSS supports a download model. A user can subscribe to an RSS feed in a reader application, Web browsers or e-mail client. The reader will download new content automatically as it becomes available to local storage and manage this content store, deleting older content as desired. This enables offline content consumption, and is well suited for mobile applications where connections may not always be available.

For connected applications, RSS provides an alternative to streaming. Content is downloaded in the background, and then played out from local storage. This enables "trickle" content distribution where the connection bandwidth is less than the media bit rate. The RSS reader effectively manages an edge cache for the user, providing instant access to high quality content, unaffected by any networking impairments due to load, packet loss, etc. Today's connected DVRs (e.g., Tivo®) and even displays (e.g. Sony Bravia® Internet Video Link®) contain feed readers and local storage to move RSS beyond the desktop to the set-top. This mechanism can offer an efficient alternative to broadcast distribution of serial television content, particularly niche content, reserving the high performance IPTV networks for live content such as sports. Basically, any content that a user watches from a DVR can be delivered via RSS or other managed download at higher quality and at much lower network engineering cost (no real-time quality of service guarantees are required.) The only downside is delay – the user must identify in advance, to which content they are interested in subscribing. Although the typical RSS feed uses HTTP over TCP to transport the media, it is possible to use peer-to-peer (P2P) content distribution, in which case the origin URL refers to a torrent seed, for example.

The RSS 2.0 XML syntax [Win03] is easy for developers and content creators to use and the typical high-level metadata of title, date, description are readily available (see Fig. 2.10 for example.) In addition to the media, RSS includes the specification of a channel icon to represent the content in user interfaces. The XML namespace mechanism allows RSS content descriptions to support additional applications and metadata such as geospatial coordinates such as GeoRSS and W3C Geo or traditional bibliographic metadata such as the Dublin Core (see Table 2.5.) Unfortunately this extensibility has led to some added complexity, incompatibilities and redundancy in the metadata specifications in use. The Atom format was proposed as an improvement and partial solution to address these incompatibilities, but until such time as RSS sunsets, the result is yet another syndication format on the landscape.

```
<rss version="2.0">
  <channel>
    <title>Rocketboom</title>
    <link>http://www.rocketboom.com/vlog/</link>
    <description>Daily with Joanne Colan</description>
    <copyright>Copyright 2007</copyright>
    <lastBuildDate>Thu, 08 Mar 2007 09:34:15 -0500</lastBuildDate>
    <generator>http://www.movabletype.org/?v=3.33</generator>
    <item>
      <enclosure
url="http://www.rocketboom.net/video/rb_07_mar_06.MP4" length="10"
type="video/mp4"/>
      <title>rb_07_mar_06</title>
      <description>story links: rb field correspondent bre pettis,
prepares for space...</description>
<link>http://www.rocketboom.com/vlog/archives/2007/03/rb_07_mar_06
.html</link>
<guid>http://www.rocketboom.com/vlog/archives/2007/03/rb_07_mar_06
.html</guid>
      <category>daily</category>
      <pubDate>Tue, 06 Mar 2007 12:44:12 -0500</pubDate>
    </item>
    <item>(additional items)</item>
  </channel>
</rss>
```

Fig. 2.10. An RSS 2.0 sample with MPEG-4 enclosures intended for the Sony PlayStation Portable (Rocketboom).

Table 2.5. A stack view of RSS protocols.

iTunes	MediaRSS	DCTerms	etc.
RSS 2.0			
XML 1.0			
Encoding (e.g. UTF-8)			

2.4.3 Podcasts

RSS and Atom include enclosure tags to refer to other media besides text, and Apple has chosen RSS 2.0 for their Podcast format (Fig. 2.11.) The widespread success of iTunes and the iPod® personal media player has resulted in the unprecedented deployment of easy to use media download management capabilities. While other systems for organizing personal media and other personal media players predated Apple's iTunes®, the single vendor environment has enabled a reliable, consistent platform for content delivery to mobile devices via download. As such, to reach the large audience of iPod® owners, content publishers have rushed to make their content available in the form of Podcasts. The standard is open and has been implemented by many consumer electronics device manufactures. For example, the Sony PSP® includes a WiFi interface and supports automated unattended syncing eliminating the requirement that users dock and sync their devices to get new content.

```
<item>
  <title>MLB Radio Daily: 04/11/2006</title>
  <itunes:author>MLB.com</itunes:author>
  <itunes:subtitle>Braves closer Chris Reitsma, Tim Brown of the
  LA Times, Jeff Blair of the Toronto Globe and
  Mail</itunes:subtitle>
  <itunes:summary>Braves Closer Chris Reitsma joins the show. Jeff
  Blair of the Toronto Globe and Mail talks about the teams expec-
  tations and how the Jays stack up in the division. Plus, LA
  Times columnist Tim Brown on the latest Gagne news and the lat-
  est improvements to the ballpark.</itunes:summary>
  <description>Braves Closer Chris Reitsma joins the show. Jeff
  Blair of the Toronto Globe and Mail talks about the teams expec-
  tations and how the Jays stack up in the division. Plus, LA
  Times columnist Tim Brown on the latest Gagne news and the lat-
  est improvements to the ballpark.</description>
  <guid>http://dds.mlb.com/mp3/mlbr_daily/041106_mlbr.mp3</guid>
 <enclosure
  url="http://dds.mlb.com/mp3/mlbr_daily/041106_mlbr.mp3"
  length="21614875" type="audio/mpeg" />
  <itunes:duration>30:01</itunes:duration>
  <pubDate>Tue, 11 Apr 2006 18:24:00 EDT</pubDate>
  <itunes:category text="Sports" />
  <category>Sports</category>
  <itunes:keywords>MLB Radio Daily</itunes:keywords>
</item>
```

Fig. 2.11. A segment showing metadata from an RSS 2.0 Podcast.

2.4.4 RSS for Content Ingest

Podcast content is typically free of DRM and uses open standards such as HTTP, XML and MP3 – these attributes, combined with the metadata and scheduled update support provide for near-ideal conditions for media search engines to ingest the content. The RSS descriptions offer great efficiencies for spiders over traditional crawling. While engines may not have the rights to store and redistribute the media streams, it is widely accepted for search engines to provide indices and direct users back to the origin. RSS feeds can point to a large collection of archived serial content, so search engines can quickly ingest and create indexed archives going back into the past in a controlled manner. The feed organizes the collection of media files into a cohesive unit with common metadata for the series. A "crawler" (actually a "feed reader") can download the small XML description and determine if there is any new content to download and ingest for indexing – there is no need to hunt around a directory tree searching for new media files. Also since RSS items include the publication time, search engines can make informed estimates of the next time that content will be published and only check for new content at that time. (RSS includes a "time to live" parameter indicating the maximum cache time, but it is generally more reliable to predict the next content publication time based on the frequency of past publications using heuristics.)

Table 2.6. Supported audio and video enclosure types for common RSS feeds.

Feed	Support enclosure types (and file extensions)
Audio Podcast	audio/mpeg (.mp3), audio/x-m4a (.m4a)
Video Podcast	video/mp4 (.mp4), video/x-m4v (.m4v), video/quicktime (.mov)
MediaRSS	any

2.4.5 MediaRSS

As Table 2.6 indicates, there are many formats that cannot be included in standard Podcasts. Also, it is common practice for sites to offer content in multiple formats and multiple bitrates. Yahoo's MediaRSS addresses some of these shortcomings; in particular, multiple enclosures are supported to offer different representations (different formats, bitrates) of the content. Yahoo's video search engine suggests that content providers use MediaRSS to publish their media for ingestion by the search engine. The MediaRSS specification goes beyond global metadata to include elements with media timestamps. This capability allows for multiple thumbnail images and text that includes a temporal component to support captions.

The simplicity combined with extensibility of RSS has resulted in widespread adoption of this format on the Internet. They are used for amateur (blogs) and professional content (e.g. TV news clips or radio programs).

2.5 Metadata for Broadcast Television

Turning now from IP video sources to the broadcast world, we find a different range of metadata standards and systems. These systems have been developed to deliver television over a range of distribution channels such as cable, direct broadcast satellite, or over the air. They support EPG and systems information as well as services such as emergency alerting, closed captioning, and content advisory for parental control (V-Chip) as required by the FCC. We will mention legacy analog standards because elements of these persist even as we have largely moved to digital TV distribution standards.

2.5.1 Electronic Programming Guide (EPG)

Interactive Program Guides (IPG) are an integral part of digital television systems, allowing viewers to choose which programs to watch and, for DVRs, which to schedule for recording in the future. The term Electronic Program Guide (EPG) is often used synonymously with IPG, but the latter implies an end user application for searching and browsing in addition to the program data itself. More recently, terms including Electronic Service Guide (ESG), Electronic Content Guide (ECG) and Electronic Media Guide (EMG) have emerged to indicate that material beyond traditional television channels may be described as well. As the number of available channels increases, the IPG becomes an invaluable tool since users can no longer remember what channel their favorite programs are on, or for that matter, which channel number is assigned to a particular broadcaster. The guide information is also used for masking the channel change delay inherent in most digital TV systems. As the user changes channels, the program title and descriptive information from the EPG can be instantly displayed along with the channel number and name; the video will begin to display a few seconds later.

Program schedule data is typically displayed as a two–dimensional grid with channels (or services, hence the name "ESG") on one axis and time along the other axis. On the set-top, these interfaces struggle to overcome the relatively low resolution TV display and distant viewing environment by employing scrolling mechanisms so that only a tiny fraction of the available guide data is displayed at any one time. Filtering by program genre such as "sports," "news," or "movies" is another means by which users can locate content using the IPG. Text entry for search is cumbersome at best using an infrared remote control, but it is an option and is typically implemented using a displayed keyboard (a.k.a. "soft keyboard" or "soft keypad") navigated by arrow keys.

IPGs are becoming commonplace on the Web (Yahoo, MSN, TVGuide, etc.) where text entry and screen resolution are less of a problem. However the instant gratification of viewing desired content such as a movie from a comfortable sitting position is absent. These IPGs are useful for programming DVRs to record content, or for determining if there is anything interesting to watch in the evening once the work day is over. As more and more users consume video content on laptops and desktops, the transition from guide browsing to content consumption will be rapid and seamless.

EPG data consists of two main classes of data: (1) video program (content) metadata, and (2) scheduling data. The former answers the question "what is on?" while the latter answers "how do I find it? (in the time/channel space)." This distinction is clear in the context of movies for example. The content metadata is what might be found on the DVD jacket: title, actors, running time, rating, etc. The scheduling information tells when and to what channel viewers must tune to watch the movie. The same content may air several different times, perhaps on different channels. However this notion that the content metadata is fixed and immutable while the scheduling data is ephemeral becomes muddied in the context of news programming. Here the content may be described as "Nightly news: today's top stories" which obviously implies a temporal dependency. For 24 hour news channels which use a cyclical programming model, the content metadata becomes less valuable than the temporal scheduling data.

These EPG data sources are of great utility for video search systems, and it is up to the application to determine which data components are most valuable. For example, for archiving movies and comedy or drama series, the content metadata is most useful, while for broadcast monitoring systems, the temporal information may be just as important. These systems must support queries such as "what was on channel 13 in the New York market at 7:24pm EST on December 14th, 1994."

EPG streams are specific to a given geographic location and service distribution method. The geographic location determines to which metropolitan area or "market" the viewer belongs (in the US, these are referred to by the FCC as "defined metropolitan areas"). Each market may have its own local programming and sports blackout rules. Furthermore, viewers may get TV content from a variety of sources including over the air (OTA) analog, cable, DBS or IPTV. Providing up to date information for all of these sources is a Herculean task indeed. Due to the size of the data and the availability of accurate scheduling information, most EPG data is delivered to terminal devices (or perhaps cable head end or IPTV VHO servers) in units representing the content to be broadcast for the next two weeks. Breaking news and programming that runs over its allotted timeslot such as extra inning baseball games are often not described properly by EPG services. However, scheduling changes on the order of days can be handled by most EPG services, for example, the baseball World Series is a best of seven games played over the course of several days, but ends early if one team wins four straight games.

There are many sources of EPG data; in the US, Tribune Media Services provides EPG through Schedules Direct® (formerly "zap2it"). Notably, use of this feed is available at nominal charge for application developers and as a result numerous applications use it. Other major service providers include FYI television, Infomedia (in Europe), and Gemstar International Group Ltd, which now includes TVGuide. The EPG data are typically distributed in an XML representation as indicated by Table 2.7 and mechanisms for XML compression and segmented delivery are employed. To reduce redundancy, these formats use a keyed record schema representation. For example, information about each series is kept in a single place and assigned an identifier, and scheduled instances refer to this global information through the use of the identifier.

Table 2.7. Representative EPG XML formats.

EPG System	Source	Usage
TV-Anytime	TV-Anytime Forum	Components used in DVB, ARIB, ATSC/IIF
OpenEPG	Consumer Electronics Association®	CEA-2033
XMLTV	xmltv.org	open standard, international
XTDV	Tribune Media Services	zap2it, primarily US
GLF	Microsoft®	IPTV, proprietary

In addition to HTTP, EPG data may be delivered in other ways. For ATSC in the US, the TV Guide On Screen (also known as Guide Plus+ from Gemstar International) service uses the National Datacast Service to provide EPG data for digital television receivers. An eight day dataset is transmitted using the PBS channels.

2.5.2 Extended Data Service (XDS)

In the US, the analog TV transport used for closed captions has been used to encode program information as well. NTSC Field 2 can carry the Extended Data Service (XDS) which repeatedly sends metadata including channel call letters, program title, media duration and current play time. This service also includes the time of day which is used on terminal equipment, and this was originally appreciated by consumers since it was used for setting their VCR clocks. XDS is specified by the EIA-766 standard. Although most PBS stations use XDS, commercial stations have not widely adopted it, and it is therefore not reliable as a metadata source for program data.

2.5.3 Program and System Identifier Protocol (PSIP)

While XDS was originally developed for transmission in analog TV systems, today this information is delivered digitally using the ATSC digital television specification using MPEG transport streams with a newer protocol, PSIP, the Program and System Identifier Protocol [PSIP02, Eyer02]. PSIP defines a set of "tables" for representing data in the MPEG-2 transport stream. The time of day is carried in a System Time Table (STT), and the ratings in a Rating Region Time Table (RTT). A "virtual channel table" is used to convey the list of available channels for guides, but of particular interest for content description are the EIT or Event Information Tables. The ATSC uses the term "Event" to describe an instance of a TV program (note that the term "program" has a different meaning in MPEG parlance). EITs are recommended to be transmitted twice a second. The EIT contains start, duration, title, and optional description, content advisory data, and metadata about the closed caption and audio (not the data itself). Receiver UIs may display only the first 30 characters of the title. Descriptions may be sent using extended text messages (ETM) in extended text tables (ETT) and up to 16 days of program data may be advertised in advance.

The Digital Video Broadcasting (DVB) consortium specifies a broad range of transmission standards for terrestrial, satellite, cable and handheld applications (e.g. DVB-T, DVB-S, DVB-C, and DVB-H). Like the ATSC specifications, DVB uses MPEG-2 transport streams (TS) but the protocol for program metadata differs and is encoded in Program Specific Information (PSI) tables to include service information (DVB-SI). This is used for delivery of EPG information in DVB systems [DVB98].

2.6 Metadata for Video on Demand

2.6.1 Introduction

Video on demand (VoD) systems allow users to choose from a collection of titles and instantly begin to view the video material. These systems have been in use for many years but are growing in popularity recently with the expansion of digital cable and IPTV. Video download services also may use VoD-formatted content as source material. As shown in Fig. 2.12, content providers deliver content via aggregators, and these in turn provide packed content to VoD service providers.

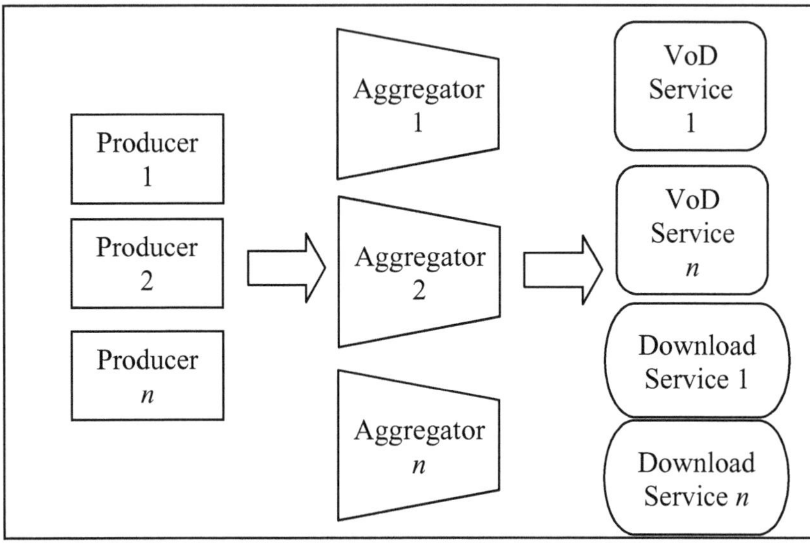

Fig. 2.12. Video on demand content flows from producers through aggregator services and on to end-user service providers.

VoD systems support playback control including pause, fast-forward, etc., often using so called "trick streams" which are separate copies of the media encoded in advance of the content being made available for users. There may be several speeds supported and the system will switch among them to simulate advancing the video at 2x, 4x, etc. realtime. To help users make a VoD selection, services provide a preview capability and content providers prepare theatrical trailers to promote their content. To enable rapid browsing of several titles in parallel, systems present box art images of the asset along with the title. Metadata for search also includes genre, rating, actors, and a short description. Media particulars such as aspect-ratio (letterbox / full screen), existence of subtitles and captions, and alternative audio languages are also available. This collection of the media itself, a preview, box art and metadata, forms a VoD "package" which is managed as a unit; for example the package has a well defined period of availability, after which the entire package is removed from the VoD servers.

2.6.2 Cable Labs

In the US, the cable companies have formed the Cable Labs, a consortium to promote standards and interoperability. Cable Labs VoD metadata and packaging standards are in widespread use for distribution of VoD content. Fig. 2.13 shows the overall structure of an ADI 1.1 package asset, while Fig. 2.14 contains a fragment of the metadata for a VoD asset in XML format. Notice that "Run_Time" has the same meaning as "total time" in Fig. 2.6.

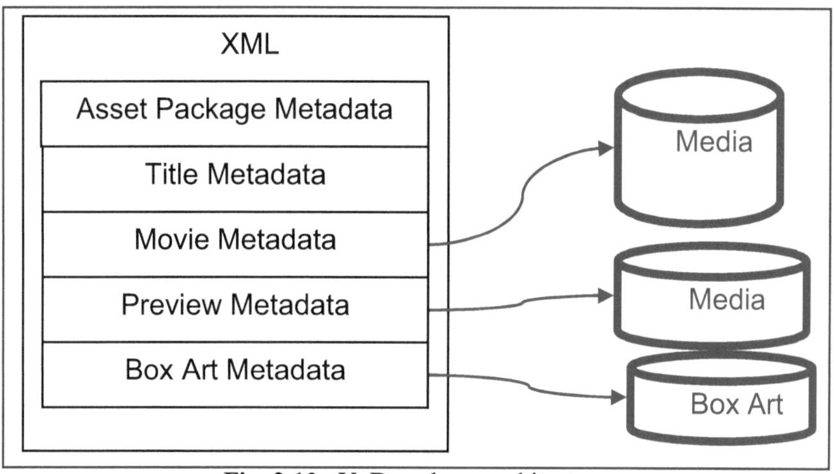

Fig. 2.13. VoD package architecture.

```
<App_Data App="MOD" Name="Title" Value="The Titanic"/>
<App_Data App="MOD" Name="ISAN" Value="1881-66C7-3420-000-7-
9F3A-02450-U"/>
<App_Data App="MOD" Name="Summary_Short" Value="Fictional
romantic tale of a rich girl and poor boy who meet on the ill-
fated voyage of the 'unsinkable' ship"/>
<App_Data App="MOD" Name="Run_Time" Value="03:14:00"/>
```

Fig. 2.14. Excerpt of CableLabs ADI 1.1 format VoD metadata.

2.7 Production Metadata

Depending on the type of production, the ratio of video footage that is shot to footage that ends up in the final production varies, but is typically in the range of 10:1 or higher. This raw content represents a great opportunity for video search systems when combined with asset management systems in the production workflow. In this section, we introduce some of the metadata standards used in professional video production systems.

AAF, the Advanced Authoring Format, is used by professional editing equipment to capture and preserve metadata during the production process. MXF, the Media eXchange Format was initiated by the ProMPEG forum as part of an effort to promote interoperability as production systems moved from legacy tape-based packaged media to file-based interchange systems. MXF received the supported of the AAF Association (now called AWMA) as well as the European Broadcast Union (EBU) and these bodies have developed and standardized MXF through the Society of Motion Picture and Television Engineers (SMPTE.) MXF's main focus is standardization for interchange of finished works (called "material packages") as well as the constituent raw footage ("file packages"), while the AAF is aimed more for production and advanced editing workflows. As a result, MXF is a subset of AAF. Also, there are other formats that have been in use prior to MXF. For example, Avid (a prominent vendor in this space) video production tools have used their own format (OMF). GXF, the Global Exchange Format, originally proposed by Grass Valley Group is used by professional producers as well. Vendors are incorporating native support for MXF, but it is envisioned that there will continue to be MXF conversion as part of the production process, due not only to content archives and assets in legacy formats, but also because MXF is not intended to meet all of the needs of every application (e.g. AAF will be used for authoring). In fact, MXF is intended as an interchange format rather than an archival storage format. A key aspect of MXF is its use of UMIDs to allow metadata to be handled separately from the "essence" media. The media can be stored in an A/V server while applications manipulate metadata or perform operations such as search and display. Of course MXF also supports inclusion of lower-level metadata such as timecodes, GPS, etc. which require frame-level precision within the media itself in addition to the header metadata. MXF is agnostic to the compression format and defines a 'generic container' for encapsulating metadata for a handful of stream formats that do not include extensible metadata support [Tahara02]. The EBU has also defined the P/Meta scheme for "metadata exchange between content producer, distributor and archive" [Hoo02]. MXF includes metadata recommended by the EBU for broadcasters called the "Geneva Scheme" metadata.

DMS-1, the Descriptive Metadata Scheme, has been defined for use with MXF and consists of three parts:

1. production Framework for global metadata;
2. the Clip Framework (collection of 'scenes' which describes a stream or track);
3. the Scene Framework for frame-level metadata and as specified in SMPTE S380M.

MXF is not XML based but uses the SMPTE 336M key-length-value format to serialize the metadata for transmission and storage although systems also use XML representations of MXF data for exchange.

2.8 Timed Text Formats

2.8.1 Introduction

We've seen how global metadata such as title and genre are represented, stored and delivered in a wide range of applications such as Podcasting and broadcast television. Formats such as MXF and MPEG-7 allow much deeper metadata specification. In particular, representation of the dialog of a program is of particular interest for video search applications, so we will focus on this aspect.

Several metadata standards include support for storing the dialog of a video program along with corresponding temporal information. For many applications such as streaming captions, DVD subtitles, and song lyrics, other specific dialog storage formats have been developed which are easier for application developers to use, even if they lack support for other basic metadata and are not extensible. This type of information is of obvious value for video search engines, and the temporal information is of importance for retrieving segments of interest in long form video. In this section, we will introduce several widely used dialog or annotation storage formats and provide examples of each. Note that, as is the case with global metadata, this information can be embedded in the media container format, or stored in separate files. The text may be synchronized or unsynchronized, and may be represented as a single unit in the header of the media file, or represented as a separate stream, multiplexed with other media streams for transport.

2.8.2 Synchronization Precision and Resolution

Time units are specified in a variety of ways such as in milliseconds relative to the start of the media, as frame numbers referencing the media time base or using SMPTE time codes. RTP defines the Normal Play Time (NPT) to represent these values. As Table 2.8 shows, applications may employ different resolutions for time stamping text segments. Note that although roll-up mode captions can be precisely specified, the timing is inaccurate due to the variable transcription delay inherent in real-time captioning.

Table 2.8. Text segmentation resolution for various applications.

Application	Text segmentation resolution
Speech recognition / synthesis	Phoneme (or sub-phoneme)
Roll-up captions	Two characters
Karaoke	Syllable
1-best ASR transcription	Word level
Music lyrics, streaming media captions	Phrase
Pop-up captions, Subtitles	Two "lines" where each line is ~30 characters
Aligned transcripts	Sentence
Distance learning, slide presentations	Paragraph

Unfortunately, as is the case with global metadata, there is no overarching agreement on the file formats or syntax for marking up timed text. The W3C is developing Timed Text to help address this problem, and the recommendation includes all possible features to support almost any application. However, it is likely that the simple, application specific text formats will persist for the foreseeable future; in fact new formats are still being created, e.g. the MediaRSS specification is relatively new, so these formats show no sign of dying out. Table 2.9 gives a sampling of some of the formats in use today to give an idea of the current state of timed text on the Web. Note that DSM-CC NPT is the Digital Storage Media Command and Control Normal Play Time which is also used by RTSP (RFC 2326 3.6).

Table 2.9. Representative timed text formats.

Format	Source	Sample
MediaRSS	Yahoo®, uses DSM-CC NPT timestamps	<media:text type="plain" lang="en" start="00:00:03.000" end="00:00:10.000"> Juncos can be black or gray</media:text>
SAMI	Microsoft®, milli-second timestamps	<sync start="3"> Juncos can be black or gray </sync>
QuickTime® Text	Apple	[00:00:08.959] Juncos can be black or gray
RealText	RealNetworks®, SMIL	<time begin="00:00:19.090"> <clear/>Juncos can be black or gray
TimedText	3GPP, MP4, W3C	<TextSample sampleTime="00:00:08.000" text="Juncos can be black or gray"> </TextSample>
Synchronized lyrics/text	ID3 Lyrics3v2 [id3 spec]	[00:11]Juncos can be black or gray[CR][LF]
Google Video	Google®	00:00:11.000 Juncos can be black or gray
SRT	One of the popular subtitle formats used in DVD transcoding	1 00:00:11,000 --> 00:00:16,400 Juncos can be black or gray

2.8.3 Transcripts

In addition to the metadata sources described above, media content in the form of text or text streams is sometimes available from the Web or other sources. Transcripts of TV program dialogs are freely available on the Web for some programs, (e.g. cnn.com/transcripts.) Others are offered through fee-based services such as Burrelles® or LexisNexis®. Transcripts usually include some level of speaker ID and may include descriptions of visuals or audio events not in the dialog (e.g. "car horn honks").

2.8.4 Closed Captions

In the US, closed captioning is used to make TV accessible to hearing–impaired viewers. Captions are textual representations of the program dialog as well as any significant audio events that a hearing–impaired viewer would need to understand the program. The term "closed" connotes that the captions can be switched on or off by the viewer. Captions are similar to subtitles used for translating films into other languages (see Table 2.10). Captions benefit all users: in certain consumption scenarios such as noisy environments, multichannel viewing or in public venues, captions are invaluable. Captions also improve comprehension for non-native listeners and can improve reading skills.

Table 2.10. Comparison of EIA-608 captioning and subtitling.

EIA-608 Captions	Subtitles
Primarily intended for hearing–impaired viewers (rarely includes "second language")	For alternative language viewers (very rarely includes cues for hearing–impaired viewers)
Character codes (modified ASCII)	Bitmapped
At most two languages	Many languages
Latin characters (approximate)	Any characters or fonts

The FCC has mandated that nearly all broadcast content must be captioned, with some exceptions for cases where it is not practical. Video services and equipment such as DVRs and DVDs are required to preserve or "pass through" the caption information, and television sets over 17 inches in diagonal are required to include the ability to display captions.

Live TV is captioned in real-time often by highly skilled stenographers and displayed in "roll-up" mode, while other productions including commercials and movies are captioned in an offline mode. In some cases, the program script for the teleprompter is fed into the closed caption. Offline captioning is displayed in a "pop-up" or "pop-on" mode where the timing can be precisely controlled (down to the frame level) and screen positioning is often used to indicate speaker changes. Real-time captioners can't afford to take the time to position text underneath the speaker, so a convention of using two chevron (greater-than) characters has been adopted to indicate speaker change. Similarly, three such characters indicate a topic change. The EIA-708 closed caption used in the ATSC DTV standard greatly expands EIA-608 and includes carriage of 608 for compatibility.

DVD subtitles are bitmaps, unlike EIA-608 captions which use a modified ASCII encoding. This provides more flexibility, but requires optical character recognition for extraction and use by search engine systems.

Many tools are available for this (subtitle rippers). Note that many DVDs include closed captions as well as subtitles.

2.8.5 Synchronized Accessible Media Interchange

Microsoft developed the Synchronized Accessible Media Interchange (SAMI) format to provide the ability to add closed captions to streaming Windows Media® format audio and video. The captions are stored in an XML-like file separate from the media but with the same root name with the extension ".smi". This often leads to confusion with the SMIL format files which may use the same extension. Although often stored in the same directory as the media, the media player can read the SAMI file from an entirely different URL and display the captions under the control of the user.

2.8.6 Metadata from Social Sources

Web users extract movie subtitles, translate, and post them on the Web (this is known as fan translation, or fansubs.) A popular movie may be translated into more than 30 languages. This phenomenon is typically in violation of copyrights and the quality cannot be assured. However, a site called DotSub is available for social multilingual captioning of Web media in Flash format, and this is welcomed by many content creators, rather than viewed as a copyright infringement. Extended character sets and encodings are needed to represent multilingual texts. Beyond dialog representation, social tagging (a.k.a. folksonomy) may be applied to video wherein any user may enter metadata which is later indexed for retrieval.

2.8.7 Metadata Issues

Practical considerations limit the reliability of authored metadata such as descriptions or keywords, particularly for non-professionally authored media. It is up to the content author to maintain this information, which is tedious and expensive, so in practice it may not be reliable. Applications such as the Windows Media® Encoder which retain the most recently used values across runs, while intended to save repeated metadata entry, may in fact cause problems. If a user is not aware of this feature, it is possible that the metadata they may end up with is more relevant to their last work, not their current one. Also, editing applications may extract metadata from the constituent clips and apply that to the edited work.

2.9 Conclusion

We've seen that there is a wide range of content sources available to video search engines, each with its own associated metadata descriptions. Metadata systems have historically arisen from distinct communities including video production, US and international television broadcasters, as well as the computing and Internet standards bodies. In each of these domains, legacy and single-vendor systems continue to play a major role, representing significant sources of described content.

For a representative sampling of systems from these domains, we've looked into metadata which describes content at a range of levels. While MPEG-7 and MXF can go much further, most systems at a minimum support specifying textual metadata with temporal attributes. Other systems such as the DCES provide only high level attributes about the media assets, but offer broad applicability. The related topic of text stream formats for capturing the dialog of video and audio media was also introduced. Although technically text streams are content as opposed to metadata, for search applications they are interpreted by retrieval systems as descriptions of the media content. Finally, metadata systems for describing collections for content such as Electronic Program Guides and RSS Feeds were shown to provide valuable metadata about their constituent media items.

References

[3GPP] 3rd Generation Partnership Project, 3GPP TS 26.244, V7.1.0, Technical Specification Group Services and System Aspects Transparent end-to-end packet switched streaming service (PSS); 3GPP file format (3GP) (2007).

[Adams06] Adams, G., Ed.: Timed Text (TT) Authoring Format 1.0 – Distribution Format Exchange Profile (DFXP), W3C Candidate Recommendation (2006).

[ASF04] Microsoft Corp., Advanced Systems Format (ASF) Specification, Revision 01.20.03 (2004).

[Burnet06] Burnett, I. et al.: *The MPEG-21 Book*, Wiley, Chichester, West Sussex, England (2006).

[DC03] "Information and Documentation – The Dublin Core metadata element set". ISO Draft International Standard 15836:2003, Feb. 26, 2003.

[DIDL01] ISO/IEC CD 21000-2:2001, Information Technology - Multimedia Framework - Part 2: Digital Item Declaration (2001).

[DMCA98] The Digital Millennium Copyright Act of 1998, US Copyright Office, Title II, Section 512 (1998).

[DVB98] Digital Video Broadcasting (DVB) – Specification for Service Information (SI) in DVB systems, EN 300 468 (1998).

[Eyer02] Eyer, M., *PSIP: Program and System Information Protocol*, McGraw-Hill, New York (2002)

[Hof02] Hoffmann, H.: File exchange formats for networked television production, *EBU Technical Review* (2002).

[Hoop02] Hooper, R.: P/Meta – Metadata Exchange Scheme, *EBU Technical Review*, No. 290 (2002).

[Kur06] Kurth, M., Basic Dublin Core Semantics, International Conference on Dublin Core and Metadata Applications in Manzanillo, Mexico (2006).

[Lag02] Lagoze, C. and Van de Sompel, H.: Open Archives Initiative Protocol for Metadata Harvesting, OAI-PMH, http://www.openarchives.org/OAI/openarchivesprotocol.html, cited June 16, 2008 (2002).

[Loomis04] Loomis, J.: *Windows Media Metadata Usage Guidelines*, Microsoft Corp. (2004).

[Lug04] Lugmayr, A. et al.: *Digital Interactive TV and Metadata: Future Broadcast Multimedia,* Springer, New York (2004).

[Nil99] Nilsson, M.: "ID3 tag version 2.3.0," Informal standard, http://www.id3.org/id3v2.3.0, ID3.org cited June 16, 2008 (1999).

[PSIP02] ATSC Recommended Practice: Program and System Information Protocol Implementation Guidelines for Broadcasters, Advanced Television Systems Committee, Doc A/69 (2002).

[QT03] QuickTime 6.3 + 3GPP, Apple Computer, Inc. (2003).

[Smi06] Smith, J.R.; Schirling, P., Metadata standards roundup, *IEEE Multimedia*, **13**(2), pp. 84–88 (2006).

[Tahara01] Tahara K., and Gaggioni, H.: *MXF: Technology Enabler for IT-Based Broadcast Operations*, Sony Corp. (2001).

[UPnP02] UPnP Forum, Content Directory:1 Service Template Version 1.01, (2002).

[Win03] Winer, D.: RSS 2.0 Specification, Berkman Center for Internet & Society, Harvard Law School (2003).

[Wit97] Witbrock, M. and Hauptmann, A.: Improving Acoustic Models by Watching Television, *AAAI Spring Symposium*, Palo Alto, CA (1997).

3 Internet Video

3.1 Introduction

Today's digital video systems can produce excellent quality visual and auditory experiences at relatively low cost. However, Internet users still encounter many problems that result in an unsatisfactory experience. Although the situation has been steadily improving, buffering delays, incompatible formats, blocky, blurry images, jerky motion, poor synchronization between audio and video are not uncommon and lead to frustration to the point that the user experience of video services involving search is greatly impacted. User's expectations are raised by their familiarity with broadcast television systems, where well defended standards, mature technologies, and abundant bandwidth prevail. In this chapter, we provide background information to shed light on the complexities involved in delivering IP video. We address the practical issues that video search engine systems must resolve in order to deliver their "product" – relevant video information – to users.

3.2 Digital Video

3.2.1 Aspect Ratio

When designing user interfaces for visualizing video search results, the frame aspect ratio (FAR) of the source video and resulting thumbnails must be taken into account. For many years the ratio of width to height for the bulk of video on the Web was 4:3, but with HD cameras dropping in price, more and more 16:9 format video is appearing. Content sourced from motion picture film may have one of several aspect ratios, but has always had a wider aspect ratio than standard definition television. It is also common to find wide aspect ratio source material digitized within a 4:3 frame in letterbox format with black bars at the top and bottom. When presenting grids of thumbnails for visual browsing, these circumstances present basic layout issues, and make the thumbnails for some content appear smaller than for others, impeding browsing.

Metadata extraction systems must accommodate video with disparate spatial resolutions. For example, a system may detect faces and represent the bounding box results in XML format for content that is 640 x 480 or 320 x 240 but render a user interface with 160 x 120 thumbnails. We can scale the thumbnails or rely on the browser to do so, but we must also scale the bounding box coordinates if we are to plot the detection results overlaid on the thumbnails using Scaleable Vector Graphics (SVG) or Vector Markup Language (VML). So any image region-based metadata must be effectively normalized for query and display to handle source images of various scales and must support different vertical and horizontal scale factors to normalize different frame aspect ratios.

Pixel aspect ratio (PAR) further complicates the matter. Early analog cameras and analog TV systems did indeed have continuous signals along the scan lines that varied in relation to the illumination – similar to the situation with audio microphones. However, in the vertical direction, the picture was sampled as is done in digital systems. There is a discrete fixed number of "lines" per frame – for NTSC we can count on 480 valid lines of picture information. Of course for digital television, we must sample in the other dimension as well, and then quantize the samples. Since the FAR for NTSC is 4:3, we should divide each line into 640 pixels so that each sample covers the same small extent of the picture in the vertical and horizontal directions – a square pixel. So why should we introduce a "rectangular pixel?" It turns out that the channel bandwidth of NTSC specification justifies sampling the signal at a higher rate to preserve image detail. 720 is commonly used and ATSC DTV also specifies a sampling resolution for standard definition video of 704 x 480. So some content may be sampled with square pixels while other content may have pixels that look like shoe boxes standing on end. A feature detector based on spatial relations (e.g. Viola / Jones) trained on square pixel data will perform poorly on rectangular pixel data, so a preprocessing image conversion step is required. Of course it is possible to scale the detector or make it invariant to scale, but this is more complex. Failure to manage the complexity of FAR and PAR correctly not only degrades metadata extraction algorithm performance, it results in objectionable geometric distortion: circles looking like ovals, and actors looking like they have put on weight.

A similar issue can arise in the temporal dimension. We may encounter video with a wide range of frame rates. Rates of 30, 29.97, 25 and 24 frames per second are common and lower bit-rate applications may use 15 f/s. Security or Webcam video may forsake smooth motion altogether and use 1 f/s to save storage. Media players can render the video at the proper rate, but motion analysis algorithms that assume a given frame rate may not perform well for all content. This effect is not usually much of a problem since the design of these algorithms intrinsically accommodates a wide range of object velocities. Think here of gait detection or vehicle counters – the absolute estimate of object velocity may be affected but the detection rate may not be.

Interlacing is another source of problems for video systems. Interlacing was introduced years ago with the first television broadcast standards to effectively double the spatial resolution given a limited bandwidth channel. The cost, however, is lower temporal resolution (and increased complexity for video processing engineers.) The frame is divided into two fields, one with the odd numbered lines and one with the even. The fields are sent sequentially transmitted. The result is fine for static pictures, but any objects that are in motion result in saw-tooth edges if the video is paused or sampled at the frame resolution. If we are subsampling to create thumbnails, this may not be a problem. The new HDTV standards perpetuate interlacing (1080i vs. 720p). The term "progressive" is used to refer to non-interlaced video, but amusingly the term "progressive JPEG" refers to something similar to interlacing. Video processing algorithms must handle interlaced sources gracefully, by de-interlacing, dropping fields, or by taking into account the slight vertical sampling offset between consecutive fields.

The relation of illumination or intensity to signal amplitude mentioned above is nonlinear and is represented as an exponential referred to as 'gamma'. Analog television systems were designed for CRTs with a nonlinear response and so precompensated the signal. Computer graphics applications and many image processing algorithms assume a linear relation.

3.2.2 Luminance and Chrominance Resolution

The human visual system cannot resolve image features that have differing hue but similar brightness as well as it can resolve features that vary in luminance. Therefore, compression and transmission systems encode chrominance information at lower spatial resolution than luminance with little apparent loss of image quality. The terms 4:2:2, 4:2:0, 4:1:1, etc. refer to the amount of subsampling of the chrominance relative to the luminance for different applications. When the image is rendered for display, it is converted from a luminance–chrominance color space such as Yuv or Y, Cr, Cb to R,G,B using a linear transform. Nonlinear transformations to spaces such as H,S,V yield a better match to the perceived visual qualities of color, but the simpler linear transformation is sufficient for coding gain. Single chip CCD or CMOS sensors designed for low cost consumer applications such as mobile phones or cameras also take these effects into account. Rather than having an equal number of R,G,B sub-pixels, a color filter array such as the Bayer checkerboard [Bayer76] is used to produce an image with relatively higher luminance resolution. This scheme has twice as many green pixels as red or blue. Another point to consider is that the spectral sensitivity of the human eye peaks in the green region of the spectrum, while silicon's sensitivity is highest in the infrared (IR). IR blocking filters are used to select the visible portion, but the sensitivity of the blue is much lower than the red. The resulting signal to noise ratio for the blue component is always lower than the green or red. Color correction processing as well as gamma correction tends to emphasize this noise. Also, color correction parameters are determined for given illumination conditions and, particularly in consumer applications, poor end-to-end color reproduction is common. Noise in the blue component, subsampled chrominance, and poor color reproduction not only degrade image quality, but also degrade performance of video processing algorithms that attempt to take advantage of color information.

3.2.3 Video Compression

Web media is compressed; users almost never encounter original, uncompressed video or audio – the sheer scale of storage and bandwidth required makes this impractical. Even QVGA resolution requires over 55 megabits per second to render in 24 bit RGB at 30 frames per second, while higher resolutions require even more bandwidth. The requirement that video be compressed has several implications for video search engine systems as we shall see.

Lossless video compression is rarely used since the bitrate reduction attainable is quite limited. Lossy compression offers impressive performance, but comes at the price of information loss – the original image or video sequence cannot be fully recovered from the compressed version. The distortion between the original the reconstructed image is often measured using the peak signal to noise ratio PSNR although this is well known to be a poor match to perceived image quality. It is extremely difficult to quantify image quality; it is highly subjective and content dependent. PSNR is an example of a "full reference" quality metric as defined by ITU-T Recommendation J.144 – "partial reference" and "no reference" techniques are used for applications where full reference data is not available, for example measuring quality at the set–top box at the end of a video delivery service [J.144]. Compression algorithms are evaluated using rate-distortion plots which reflect attempts to approach the information theoretic limits outlined in Shannon's rate distortion theory. Algorithmic improvements have made great strides in pushing the theoretic limits, while Moore's law has allowed for increasingly complex implementations to be standardized and used in practical systems.

Since video is a series of still frames, one would expect that video compression is related to the JPEG image compression used in digital cameras, and, in fact, this is indeed the case. Many consumer cameras capture video as a sequence of JPEG frames to create "Motion JPEG" (M-JPEG) format since the computational complexity of this approach is minimal. At the high end, professional editing systems use M-JPEG or "MPEG-2 I frame-only" as well. Here the systems are designed for high-quality and ease of cutting and splicing sequences together, rather than on high compression ratios.

JPEG works by dividing an image into small blocks and transforming (using the Discrete Cosine Transform) from the pixel domain to the spatial frequency domain. In this domain, pixels whose intensity values are similar to their neighbors can be efficiently represented – in smooth areas of an image, an entire block can be approximated by just its average (or DC) value or just a few DCT coefficients. To get an intuition for the concept of spatial frequency, take a look at a folder of digital photo files and sort them by the file size. The larger files will have a large proportion of the image in sharp focus with a lot of edge information, say from a brick wall or a tree with leaves. The smaller, more compressed, files will be the out of focus shots or contain a small object on a large homogenous background. Now suppose that we point a camera at a brick building and capture a video sequence in vivid detail. The frames are nearly identical – they have a high degree of temporal redundancy. By subtracting the second frame from the first, we end up with a frame that is mostly uniform, perhaps with a small region where someone sitting by a window in the building moved slightly. As we have found, this is the type of image that compresses well, so that our entire sequence can be efficiently represented by encoding the first frame (intra-frame coding) followed by encoding the difference between this frame and subsequent frames (inter-frame coding). Now of course there are some complications that arise due to temporal noise in the signal, and illumination changes due to passing clouds, etc. But the main problem in this scenario is that slight camera motion will result in a large difference image in any region where the image is not uniform (e.g. the sky will not cause much of a problem.) Video coders compensate for this using block-matching where a block of one frame is compared to several neighboring blocks in another subsequent frame to find a good match. In the case of a shift in the camera, most blocks will have the same shift (or motion vector). So, video compression from MPEG-1 up through MPEG-4 is based on DCT of motion compensated frame difference images.

Video compression standards are designed and optimized for particular applications; there is no one-size-fits-all codec. The ITU developed the H.261 and H.263 for low bitrate, low latency teleconferencing applications. For these applications, the facts that the camera is usually stationary (perhaps mounted on pan-tilt stage next to a monitor) and that conferencing applications typically involve static backgrounds with little motion greatly help improve the quality at low bitrates. It is reasonable here for coders to transmit intra-coded blocks rather than entire frames. MPEG-1 was developed for CD-ROM applications with bitrates in the 1 Mb/s range. MPEG-2 is used in broadcast distribution and in DVDs where higher quality and interlaced video support are requirements. MPEG-4 brings increased flexibility and efficiency, of course with increased complexity, and finally the ITU and MPEG bodies have achieved interoperability with MPEG-4 part 10, ITU H.264/AVC. For contribution feeds or editing applications M-JPEG or similar intra-coded video at very high bitrates is appropriate to ensure quality downstream.

MPEG-2 Systems [Info00] added a wide range of capabilities that were not available with MPEG-1. While "program streams" are used for file based applications (MPEG uses the term DSM – Digital Storage Media) which have negligible error, the notion of a transport stream was introduced to allow for efficient delivery over noisy channels such as may be found in typical broadcast systems such as cable or today's IPTV over DSL. The transport stream specification also supports multiplexing several (even independent) media streams which enables secondary audio programming or alternative representations of the video at different resolutions and bitrates [Haskell97]. Table 3.1 lists a few common video compression standards and bitrates typically encountered. For actual maximum and minimum bit rates supported, readers should consult the standard documents.

Table 3.1. Applications of video compression systems (bit rates are approximate, and assume standard definition).

Standard	Typical bitrates	Common applications
M-JPEG, JPEG2000	Wide range, up to 60M	Low cost consumer electronics, High end video editing systems
DVCAM	25M	Consumer, semi-pro, news gathering
MPEG-1	1.5M	CD-ROM multimedia
MPEG-2	4–20M	Broadcast TV, DVD
MPEG-4 / H.264	300K–12M	Mobile video, Podcasts, IPTV
H.261, H.263	64K–1M	Video Teleconferencing, Telephony

Within all of these standards, there are "profiles" which are particular parameter settings for various applications. The latter standards have a wide range of flexibility here which allows them to span a wide range of applications while the earlier standards are more constrained. So it is possible for an MPEG-4 decoder not to be able to decode an MPEG-4 bit stream (e.g. if the decoder only supports a baseline profile). Profiles are intended for varying degrees of complexity (i.e. required computational power of encoders / decoders) as well as latency or error resilience. For example, for DVD applications, variable bit rate (VBR) encoding allows bits required to represent high action scenes to be effectively borrowed from more sedate shots. Of course, the player has to read large chunks of data from the disk and store them in a local buffer in order to decode the video. On the other hand, for digital broadcast TV, rapid channel change is desirable so the buffering requirements are kept to a minimum. The quality difference between DTV and DVD leads many viewers to think that DVDs are HD while in fact only Blue Ray and HD-DVDs support higher resolution than standard definition. Some of this confusion arises because DVDs are often letterbox, but primarily it is due to the lack of obvious coding artifacts such as blocking or contouring. Higher bitrates play a role, but even at the same bitrate, real-time encoding for low latency applications results in lower quality. Additionally, the quality of the source is key – some digital television sources are of dubious quality, perhaps with multiple generations of encoding – as well as the fact that mastering DVDs is done offline, allowing for two-pass encoding. DVD mastering is really an art; a bit like making a fine wine as opposed to producing grape juice. So, encoding systems designers have a challenging job to balance latency, complexity, error resilience, and bandwidth to achieve the quality of experience that the viewer ultimately enjoys.

What implications do these video compression systems have for video search engines?

- Video content analysis / indexing algorithms must either support the formats natively, or transcode to a format that is supported. Since many algorithms operate in the pixel domain as opposed to the compressed domain, this "support" may simply imply that the system can decode the video. However, the video quality does have an effect on indexing accuracy – noise or image coding artifacts such as blocks can be significant problems. Also, in some cases, periodic quality fluctuations due to poor bit allocation between intra- and inter-coded frames can produce more subtle artifacts.
- Of course from a systems perspective, high bitrate video may not be practical to archive on-line at scale. Further, each format must be supported by the client media player, and by media servers as well. This problem of incompatible media players and formats is driving a move to Flash formats, which at least offers a degree of independence from the client operating system.
- Finally, as we have seen, these codecs are highly optimized for particular applications, and this typically does not include streaming or fine grained random access.

MPEG frames are organized as "groups of pictures" or GoP which consists of an intra-coded frame (I frame) and several predicted frames (P and B frames). Applications such as media players can't jump into a video stream in the middle of a GoP and start playing – they must refer back to the I frame. So in effect the GoP length determines the precision for media replay requests. For many applications the GoP length is less than a second (15 frames is common) so this has only minor effects on the user experience, but for high coding efficiency applications, "Long GoP" coding is used where there may be several seconds between I frames. H.264/AVC introduces many more complex options in this area such as multiple reference frames for different macroblocks which further exacerbate random access [Rich03].

3.3 Internet Protocol Media Systems

3.3.1 Transport

Video search engines deliver their product to clients over IP connections in several ways:

- Download – This simple delivery system has been available since the beginning of HTTP where MIME types are used by browsers to launch the appropriate media player after the media has been downloaded to a local file.
- Progressive Download – Again, a basic HTTP server delivers the media file, but in this case its play-out is initiated via a media player before the entire file is downloaded.
- HTTP with byte offsets – The byte range feature of HTTP/1.1 is used to support random access to media files. Clients map user play position (time seek) requests to media stream byte offsets and issue requests to the server to fetch required segments of the media file.
- Managed Download – A specially designed client application provides additional features such as DRM management, expiration, reliable download and HTTP or P2P is typically used for transport. There are many types of these applications, from applications that operated in the background without much of a UI, to iTunes which include download management capabilities for Podcasts and purchased media.
- HTTP Streaming – These systems require a dedicated media server that parses the media file to determine the bit rate and delivers the content accordingly. Random access and other features such as fast start, fast forward, etc. may also be supported.
- RTSP / RTP – A media streaming server delivers the content via UDP to avoid the overhead of TCP retransmissions. Some form of error concealment or forward error correction can be used. Some IPTV systems use a "reliable UDP" scheme where selective retransmission based on certain conditions is employed.

3.3.2 Searching VoD vs. Live

Most video search applications inherently provide personalized access to stored media – essentially this is a "video on demand" (VoD) scenario, although the term VoD is commonly used to refer to movie rental on a set-top box delivered via cable TV or IPTV. For VoD, the connection is point to point and unicast IP transmission is appropriate. However, IPTV and Internet TV are channel based where many users are viewing the same content at the same time so multicast IP is employed. As the number of these feeds grows, users will need searching systems to locate channels of interest. In this scenario, EPG/ESG data including descriptions will provide the most readily accessible metadata for search. Live streams can be processed in real time to extract up to the minute metadata for more detailed content-based retrieval. Of course, prepared programming and re-broadcasts of live events can be indexed a priori and used to provide users with more accurate content selection capabilities.

3.3.3 IPTV

IPTV is often heralded as the future of television, promising a revolution on the same scale as the Web. With all this potential, there are many groups co-opting the term IPTV to their own advantage. Does IPTV imply any television content delivered over an IP network? Well, we have been able to see video content streamed over the Internet for years so it makes sense to restrict the term IPTV to a narrower connotation. Of course, as more bandwidth has become available and desktop computers more powerful, we can experience full-screen video delivery and begin to approach broadcast TV quality (although HD delivery to large audiences over unmanaged networks is much more demanding and may be slow to evolve). The term "Internet TV" has been used to describe this type of system, and the term IPTV is generally accepted to mean delivery of a television-like experience over a managed IP network. To avoid confusion for the purposes of standardization, the IPTV Interoperability Forum (IIF) group formed by the Alliance for Telecommunications Industry Solutions (ATIS) [ATIS06] has defined IPTV as:

> the secure and reliable delivery to subscribers of entertainment video and related services. These services may include, for example, Live TV, Video On Demand (VOD) and Interactive TV (iTV). These services are delivered across an access agnostic, packet switched network that employs the IP protocol to transport the audio, video and control signals. In contrast to video over the public Internet, with IPTV deployments, network security and performance are tightly managed to ensure a superior entertainment experience, resulting in a compelling business environment for content providers, advertisers and customers alike.

In the context of video search, IPTV is a significant step towards an evolved state of video programming where the entire end-to-end process is manageable using generic IT methods. While there is clearly a long way to go in terms of interoperability and standardization for exchange of media and metadata, the IP and accessible nature of the new delivery paradigm paves the way toward making this a reality. This offers the potential for engineers competent in networking and data management technologies to bring their experience to bear on the problem of managing video distribution. The potential for metadata loss through conversions through the delivery chain is greatly reduced. Of course, today's IPTV systems use IP for distribution to consumers, but IP is not necessarily used for contribution of broadcast content. Traditional and reliable methods used for cable delivery such as satellite, pitcher / catcher VoD systems, etc. will persist for the foreseeable future. In addition to ATIS, several other bodies including ETSI (DVB-IPTV), OMA (BCAST) and OpenIPTV are participating in drafting IPTV recommendations for a range of applications.

Although not specified in the ATIS/IIF definition, IPTV deployments are usually delivered via DSL links that do not have enough bandwidth to support the cable model of bringing all channels to the customer premises and tuning at the set-top. With VDSL2, downstream bandwidth is typically 25Mb/s which can accommodate two HD and two SD channels simultaneously. With IPTV over DSL, only a single channel for each receiver is delivered to the customer – effectively the "tuning" takes place at the central office. This is sometimes referred to as a "switched video" service (although the term is used is used in cable TV delivery as well). To support rapid channel changing, IPTV systems keep the GoP short and employ various techniques to speed up channel change. Of course short GoP and channel change bursts consume bandwidth and systems must balance these factors. Given this optimization, and the necessary FEC for DSL, IPTV streams must be transcoded for efficient archival applications where there is less need for error correction.

As we have seen, there are a wide range of video coding systems in use and each is optimized for its intended set of applications. As video content is acquired and ingested into a video search engine, it is very likely that the encoding of the source video is not appropriate for delivery from the search engine. In some cases the bit rate is simply too high to scale well given the number of concurrent users, or the format may be unsuitable for the intended delivery mechanism. Although some services attempt to redirect users to origin servers, the user experience of switching among multiple players (some of which may not be installed) to view the search results is less than seamless. Therefore many systems have opted to transcode video to a common format and host it. Flash Video is often the format of choice here due to its platform independence and wide installed base of players. The term transcoding is loosely used to refer to changing container formats, encoding systems, or bitrates. Transrating refers to changing only the bitrate (typically via re-encoding, not using scalable coding or multirate streaming). In some cases it is not necessary to fully decode the media streams and re-encode them, such as when changing only the container format. Also, the re-encoding process can be made more efficient by only partially decoding the source (perhaps re-using motion estimation results), but in many general purpose transcoding systems, the source is fully decoded and the results fed to a standard encoder. This approach is taken because the required decoders and encoders are readily available and have been highly optimized to perform efficiently. Also, search engines may transcode to a small set of formats in order to target different markets such as mobile devices (e.g. YouTube's use of Flash Video required large scale transcoding in order to support AppleTV® and iPod Touch® which did not include support for Flash Video).

3.3.4 Rights Management

In addition to incompatible media formats, digital rights management (DRM) systems are not interchangeable, and systems that hope to process a cornucopia of content must navigate these systems as well. Various DRM systems such as Apple's FairPlay and Real's Helix can be applied to MPEG-4 AAC media, but this does not imply interoperability. While it would be in keeping with the sprit of DRM to allow the purchaser of a song (or a license to a song) to enjoy the media and justly compensate the provider, in practice this notion has been restricted so that the user must enjoy the song on a single vendor's device or player. MPEG-21 attempts to standardize the intent, if not the particular implementation, of rights through the definition of a rights expression language (REL). Examples of limited rights to use content include play once, play for a limited time, hold for up to 30 days and then play many times for up to 24 hours after the first play. The hope is that at least these desired use cases can be codified even though a particular media player device may only support a limited number of DRM systems or only a single system. In reality, choosing a DRM system is tantamount to choosing a media player. Purchased music and media (iTunes, Windows Media), video download services, DVDs and broadcast television all have forms of encryption for prevention of unauthorized copy of content (CCS, AACS for DVDs, conditional access for DVB and Cable, broadcast flag for ATSC). Finally, media watermarking and embedding user information in metadata to enable forensic traceability of a copied asset to its source are additional techniques used to preserve the copyright owner's rights.

3.3.5 Redirector Files

Video search engine systems can make use of redirector (or "metafiles") to provide increased functionality when initiating video playback. Instead of the user interface containing links directly to the media files, the links point to media metafiles which are small text markup files issued by the HTTP server with a particular MIME type that is mapped to the client media player. At this point the browser has done its job and control of the streaming session is passed to the media player which connects to a media server. This arrangement provides several advantages:

- Response time: the small files download instantly and the media player application can launch quickly and begin video playback using progressive download or streaming.
- Failover / Loadbalancing: The redirector files can include alternative URLs for retrieving the media and media players support a failover mechanism where connection to servers indicated by a list of URLs is attempted in sequence. Applications can also generate metafiles dynamically with URLs pointing to lightly loaded streaming servers if the desired media is avalible on multiple media servers.
- Playtime offsets / clipping: the media play time start and duration can be encoded in the metafile. The ability to seek into the media is critical for directing users to relevant segments in long-form content.
- Playlists / Ad insertion: sets of media files matching user quieres can be represented as a play list and interfaces supported by the media player can be used to navigate amoung them. Preroll or interstitial advertizing can be supported using this mechanism – where essentially one or more clips in the playlist are ads. Much to users chagrin, these clips can be marked so that the ability to skip or fastforward are disabled during playback of ads.
- Additional features: Optionally, directives for including media captions (similar to closed captions) are supported. Also, metadata specific to the session can be included, e.g. the title can be set to "Results for your query for the term: NASA." This mechanism can be used to effectivly override any metadata embedded in the media itself.

Table 3.2. Media metafile systems.

Format	Extension	Comments
Real Audio Metafile	.ram	One of the early streaming Web media formats
Windows Media Metafile	.asx, .wmx, .wvx, .wax	Extensions connote video (v), and audio (a) but the format is the same; 'asx' is deprecated
Synchronized Media Information Language	.smil, .smi	Supports many additional features such as layout.

Some common file formats or protocols for achieving this effect are shown in

Table 3.2; also the playlist formats such as M3U and PLS provide a somewhat similar function, but with a limited subset of the capabilities.

Fig. 3.1 shows a Windows Media Format metafile that includes failover (if the media is not available from mserver1, then mserver2 will be contacted). Also the media play position is set to 120 seconds. For Quicktime, a “reference movie” can be created to point to different bitrate versions of the content. A Reference Movie Atom (rmra) can contain multiple Reference movie descriptor atoms (rmda).

```
<ASX version = "3.0">
 <Entry>
  <Ref href="http://mserver1.company.com/media/video1.wmv"/>
  <Ref href="http://mserver2.company.com/media/video1.wmv"/>
  <StartTime Value="120"/>
 </Entry>
</ASX>
```

Fig. 3.1. ASX Metafile with failover and start offset.

Embedded players: While UIs that launch the media player using metafiles can be extremely lightweight (no client side JavaScript is required) and therefore easily supported by a wide range of browser clients, a more integrated user experience is achieved by embedding the media player in the browser. With this approach, the player plug-in loads once and user navigation of search results can change the media and change the play position. For example Fig. 3.2 shows a client side script fragment for loading a media stream and seeking to a given point using the Windows Media Player object model, assuming that the player has been embedded and named “Player”. More recently, immersive interfaces that provide a user experience more similar to TV have been created leveraging emerging technologies including AJAX, XAML, and using the graphics capabilities of clients to their full potential to provide full screen interfaces with overlaid navigational elements.

```
Player.URL = "http://server1.company.com/media/video1.wmv";
Player.controls.currentPosition = 120;
Player.controls.Play();
```

Fig. 3.2. Controlling media playback using client side scripting.

3.3.6 Layered Encoding

Some encoding systems include features to efficiently support scalability. Scalability encompasses several varieties including spatial, temporal, and even object scalability. The idea is to encode media once and enable multiple applications where views may be alternatively rendered for services with bandwidths less than the media encoded bitrate. The concept also supports the notion of a base layer and an enhancement layer where the base layer may represent a lower resolution or lower frame rate version of the media and the enhancement later can include more detail. In a best effort network delivery scenario with variable congestion, the base layer can be delivered with a guaranteed quality of service (QoS), while the enhancement layer can use a lower priority so that the overall system user experience will be improved. (Rather than one user – or worse all users – experiencing video dropouts, all users may see a slight degradation in quality).

Some media streaming systems use a less efficient scheme to provide a similar effect. Using what is called "multirate encoding," multiple versions of a video encoded at different bitrates are merged into a single file. Some implementations of this can be very inefficient, in that each stream is self-contained and doesn't share any information from the other representations of the media. Streaming media players can detect connection bandwidth dynamically and switch among the streams as appropriate. While crude, this improves the situation over the case where the user must select from separate files based on their connection bandwidth. Most users don't have a good understanding of their connection bandwidth in the first place and requiring a selection choice is poor system design which can lead to errors if the wrong setting is chosen.

3.3.7 Illustrated Audio

Illustrated audio is a class of content that fills the gap between full motion video and a bare audio stream. There are two main classes of this; the first is frame flipping where a single still image is displayed at a given point in time until the next event where a different frame is displayed. This can be thought of as non-uniform sampling: instead of each frame being displayed for the same amount of time, e.g. 33 ms, a frame may be displayed for 20 seconds followed by a frame displayed for 65 seconds, etc. An example of this is a recording of a lecture containing slides. The second class involves some form of gradual transition between slides and may include synthesized camera operations such as panning and zooming. Some replay systems for digital photographs employ this technique using automatically selected operations. Readers may also be familiar with the historical documentary style of Ken Burns where old photographs are seemingly brought to life through appropriate narration and synthesized camera operations [Burns07].

The value of this form of content has justified the creation of systems designed for efficiently compressing and representing this unique material. Microsoft's Photo Story application allows manual creation of slide shows from still frames and encodes the result in Windows media format with a special codec. Alternatively, the Windows Media Format allows for synchronous events to be included in the stream which may include links to images or may encode generic events that can be accessed via client JavaScript at media replay time to take action (which may also include fetching an image from a URL and displaying it).

Apple's Enhanced Podcasts are MPEG-4 files with streams containing specific information that allows for the inclusion and synchronized replay of embedded still images (as well as other information such as links) that can be replayed on iPods. These files typically have the extension .m4a or .m4b. The points in the media where the images are inserted naturally form waypoints for navigating in the content, and Apple emphasizes this by referring to these points as chapter markers and exposing this up through the user interface of iTunes® and iPods®. Other formats also support chapter metadata such as ID3v2 which specifies CHAP (Chapter) and CTOC (Table of Contents) and the DVD specification. In Flash video, the "Cue Point" mechanism is used for synchronizing loading of graphics and providing for navigation of the media.

For video search engines, textual chapter metadata can augment the global metadata and can improve relevance ranking and navigation for systems that support navigating within long form content. Additionally, where archiving systems manage wide varieties of content and adapt it to produce content for consumption scenarios where the primary media track is audio (i.e. mobile listening), the ability to automatically insert chapter markings to aid user navigation is extremely valuable.

3.4 Media Captioning

We have already seen how captioning can be exploited for video search, but further, video search engine systems and IP media systems should preserve any captioning that accompanies the ingested source media in order to reach the broadest possible audience. Again, it is important to point out that captioning is not just for the hearing impaired, but can improve comprehension and enable media consumption in a wider range of environments (e.g. meetings). Most IP media formats support some form of timed text and these were covered in detail in Chapter 2. The National Center for Accessible Media at WGBH pioneered television captioning [Robson97] and has recently formed the Internet Captioning Forum with industry leaders. The Distribution Format Exchange Profile (DFXP) is a subset of the Timed Text Authoring format intended to aid in interoperability of existing legacy formats. While its scope is limited, the specification includes enough generality to support a very wide range of existing captioning presentations (perhaps only exclusive of sign language representations) so it is not trivial by any means [TT06].

3.5 Conclusion

We have presented many of the practical aspects of digital video that content-based video search engine systems must deal with in order to operate seamlessly on a wide variety of content sources. At the basic level, issues of encoding and container file formats, and DRM systems must be taken into account in the system design. Next, presentation issues such as aspect ratio and transcoding for archival storage and delivery for a range of applications must be considered in the design of user interfaces for search. We also introduced methods for creating networked user interfaces for media replay with thin clients such as media players with dynamically generated playlists or browser plug-ins. Beyond the basic input and output media handling and rendering, systems that operate on the video content must also deal with real-world issues such as subsampled, noisy chrominance, non-square pixels and various temporal sampling rates. While a theoretician might correctly dismiss many of these issues as engineering decisions arising from legacy (or worse, commercially motivated proprietary and incompatible) implementations, some are related to basic principles or physical properties. There are limits to the fractional bits per pixel to which video can be compressed and the signal to noise ratio of imaging devices.

References

[ATIS06] Status Report on the work of the ATIS ITPV Interoperability Forum (IIF), ATIS / ITU, Document 34-E, Geneva (2006).

[Apple07] QuickTime File Format Specification, Apple Computer, Inc. (2007).

[Rich03] Richardson, I.: *H.264 and MPEG-4 Video Compression: Video Coding for Next-generation Multimedia*, Wiley, Chichester, West Sussex, England (2003).

[Bayer76] US Patent 3,971,065 Bryce E. Bayer Color imaging array, July 20, 1976.

[Haskell07] Haskell, B. et al.: *Digital Video: An Introduction to MPEG-2,* Chapman & Hall, New York (1997).

[Info00] Information Technology – Generic Coding of Moving Pictures and Associated Audio Information: Systems, ISO/IEC, International Standard 13818-1, 2nd ed., December 1, 2000.

[Burns07] Burns, K.: Museum of Broadcast Television article, http://www.museum.tv/archives/etv/B/htmlB/burnsken/burnsken.htm Encyclopedia of TV, 2nd ed., cited 11 January 2007.

[TT06] Timed Text (TT) Authoring Format 1.0 – Distribution Format Exchange Profile (DFXP) W3C Recommendation, November 16, 2006.

[Robson97] Robson, G., *Inside Captioning,* Cyberdawg Publishing (1997).

4 Video Search Engine Systems

4.1 Introduction

All search engine systems share a common architecture at a high level, but vary widely depending on the application and design choices. In general, there are three main architectural components as we view the system from a content flow perspective: content acquisition, processing (indexing), and retrieval (see Fig. 4.1). In practice, these are typically decoupled independent processes in order to ease scaling. We will also consider the system from a user activity perspective in which we can consider behaviors and system states.

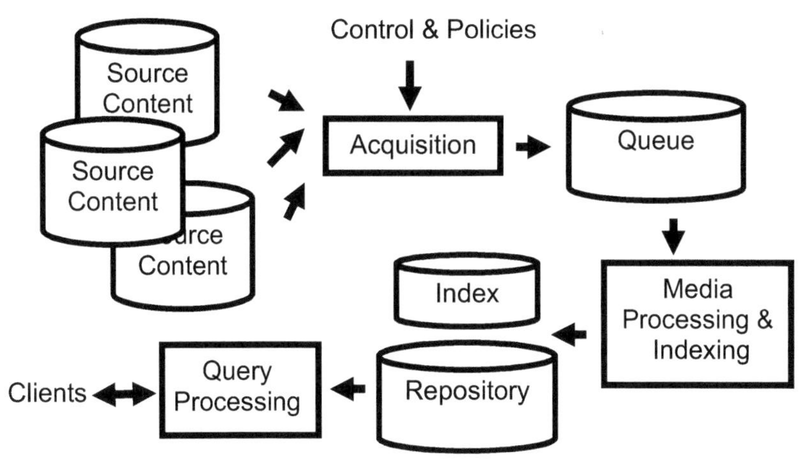

Fig. 4.1. High level architecture of a typical video search engine.

Acquisition refers to bringing source video content into the system and positioning it for subsequent indexing. This may involve copying the bulk media to local storage as in a traditional text search engine, or other modalities such as user contribution or even capture of live feeds. Acquisition is constrained or configured; for example, a list of content sources or RSS feeds may be used. Content providers may use the Outline Processor Markup Language (OPML) to create lists and publish them to search engines. Even in the case of a general Web crawl, the prior state of the crawl is used to direct future content location attempts, so the process is not free of constraints. Efficient crawling is well studied [Cha03] and like other aspects of scalable search, is typically implemented in a distributed fashion.

Content or media processing is the next logical stage in the content flow and involves transcoding, metadata manipulation, extraction and augmentation through media analysis methods. The goal is to capture the media structure and metadata in data structures that enable rapid retrieval and content adaptation.

The third major functional block from a content flow perspective is retrieval where a query engine responds to user requests in a real-time interactive mode. The results are exposed through one or more user interfaces and multimedia summaries or contextual information may be generated to improve the user experience. In addition to real-tine query handling, modules for personalization or data mining can operate on the stored multimedia collection in an offline fashion to produce customized views or analytical results for users.

4.2 Content Acquisition

4.2.1 Metadata Normalization

The acquisition module typically performs some degree of data normalization, although this can be deferred to the indexing module. The goal of metadata normalization is to simplify the rest of the system by mapping tags with similar semantics to a single tag. Unfortunately, for many systems this is a lossy process since nuances in the source metadata may not be preserved accurately. For example, the search engine may list all results with their title, but for some content, a subtitle or episode title may be included in addition to the main title. Either this subtitle information is dropped, or somehow mapped into another supported field, perhaps using a tag thesaurus, such as the content description field – or some convention may be adopted to concatenate the subtitle onto the main title. All of these alternatives generate undesirable consequences – either the field in question is not available for search, or in the concatenation case, the original tag may not be recoverable from the database. The ideal situation is to preserve all source metadata, while extracting the reliable common tags such as the DCMI fields and utilizing a searching and browsing architecture that can seamlessly manage the complexity of assets having varying numbers and types of metadata fields – a daunting task to achieve at scale. Many systems in use on the Internet today fall short of the ideal case, but still provide an adequate user experience.

4.2.2 User Contributed

Content acquisition for the user contributed case is shown in Fig. 4.2. Here, a Web form is used to capture metadata from the user, and the media is transferred up to the server. The upload file transfer may be simple HTTP, or a special purpose client application maybe utilized which provides additional benefits such as parallel uploading, ability to restart aborted partial transfers, etc. Note that it seems logical given limited available upload bandwidth to transcode source media down to a smaller size prior to uploading. However, most systems designed for Web users do not employ client-side transcoding for a number of reasons including:

1. the desire to keep the client as thin as possible to ease maintenance;
2. the desire to support clients with limited compute power;
3. possible license issues with codecs;
4. the assumption that video clips will be short, limiting overall file upload size.

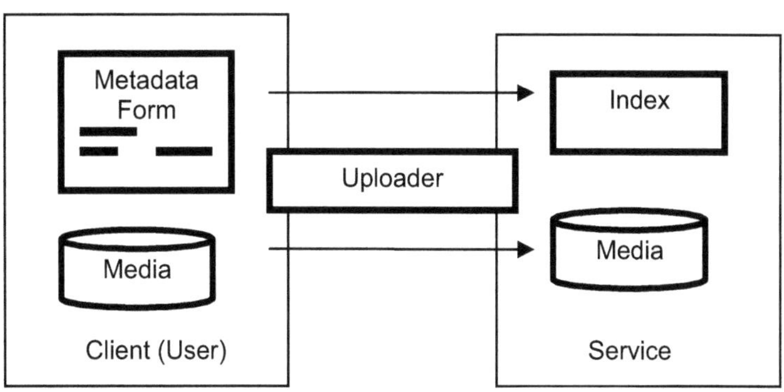

Fig. 4.2. Data flows for uploading user contibuted content.

The metadata entry may be decoupled such that the content is uploaded, and metadata tags added later either by the original author, or by other users (social tagging). Consider here the example of mobile video capture and share. The service can be such that the capture format is supported for upload (avoiding the necessity for transcoding on the mobile device), and users could access their clips from a laptop where annotating, forwarding to friends, etc. is much easier given the powerful user interface capabilities of the laptop as compared to the cellular handset. Contributors control the publication of, and rights to use (view or download), the content using categories such as:

1. Public: available to all users;
2. Groups: viewable to selected users;
3. Unlisted: unpublished; a URL is returned to access the content.

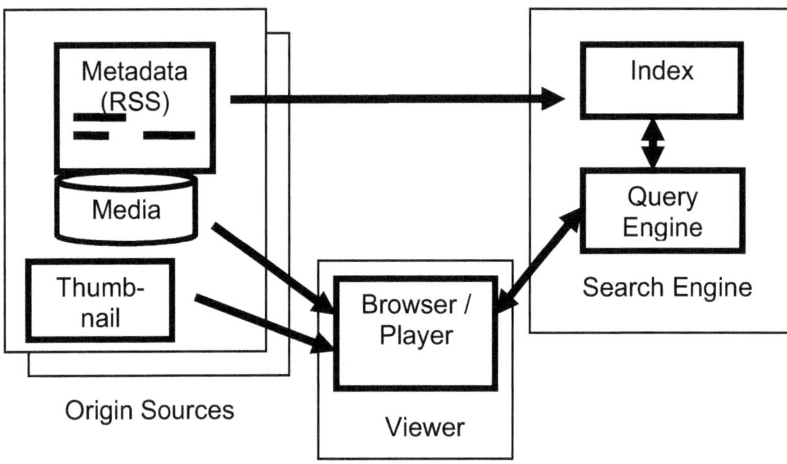

Fig. 4.3. Metadata and media flows for Podcast aggregation search.

4.2.3 Syndicated Contribution

Although we mentioned embedded metadata and introduced the concept of asset packages, which represent media and metadata as a logical unit, it is usually the case that the metadata and the media take separate paths through the acquisition, processing and retrieval flow. Perhaps the most striking example of this is the case of Podcast aggregation search sites. As shown in Fig. 4.3, it is not even required that the server ingest a copy of the media, the system can operate using only metadata provided by the source. In this case, the concept of acquisition really consists of creating a new record in the database with an identifier of some kind, perhaps a URI, referring back to the content source. It is most common to use RSS with media enclosures for the metadata formatting, with the iTunes® extensions for Podcasts or MediaRSS extensions for increased flexibility. Of course, our focus is on media processing to augment the existing metadata to enable content based search, in which case access to the media is required.

With distributed (meta) search, there is even less centralization and data movement. Rather than querying content sources for lists of new content and its associated metadata periodically and building a centralized index, metasearch systems can distribute the queries to several search systems and aggregate results. The problem of merging ranked results from disparate sources as well as duplicate removal present themselves in these scenarios.

4.2.4 Broadcast Acquisition

In addition to crawling, syndication, and user contribution, other forms of acquisition include broadcast capture and event based capture. Broadcast capture may involve analog to digital conversion and encoding, but it is becoming more common to simply capture digital streams directly to disk. For consumers, broadcast may be received over the air, via cable, via direct broadcast satellite, through IPTV or Internet TV multicast. Event-based acquisition is used in security applications where real-time processing may detect potential points of interest using video motion detection and this is used to control the recording for later forensic analysis. It is in the context of these streaming (or live, linear) sources that real-time processing considerations and highly available systems are paramount. In the previous examples of acquisition that we were considering, there is effectively a source buffer so that if the acquisition system were to go offline the result would only be slightly delayed appearance of new content – for streaming acquisition, this would result in irretrievable content loss. Of course, for user contributed situations high reliability is also desired to preserve a satisfying user experience.

A particular set of concerns arise with user contributed content, and appropriate mitigation steps should be taken prior to inserting UGC into the search engine for distribution. Users may post copyrighted material, inappropriate or offensive content and may intentionally misrepresent the content with inaccurate metadata. Sites may employ a review process where the content is not posted until approved, or may rely on other viewers to flag content for review. Fingerprinting technology is used to identify a particular segment of media based on features, and this operation can take place during the media processing phase to reject posting of copyrighted content.

4.3 Content Processing

Newly acquired media is available in a state that facilitates subsequent operations on the media such as transcoding and metadata extraction. Transcoding engines may operate independently on the content with the goal of preparing the content for delivery (perhaps via streaming) and normalizing to a single format suitable for archiving. We can consider a path for the media separate from the path for the metadata. The media will be positioned on user-facing media servers or origin servers for content distribution. Metadata will head to the index and storage for use in browsing (Fig. 4.4) and be tied together using content identifiers.

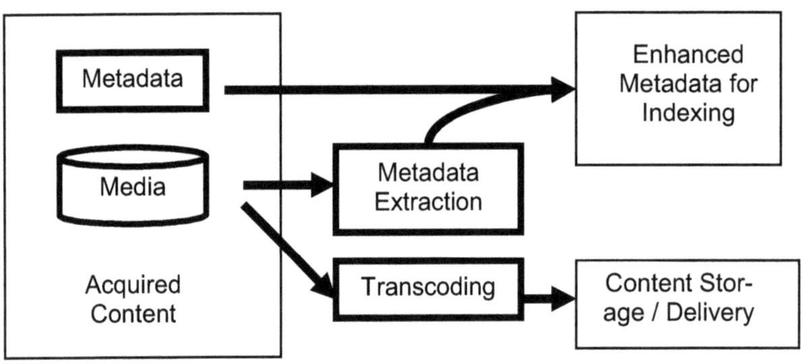

Fig. 4.4. Metadata augmentation via automated content processing.

4.3.1 Asset Management

We've presented a one-directional data flow model for content ingest and processing, but we did allude to other alternatives. Certainly, social tagging is architected such that the asset is not simply posted into the database and then read out for consumers. The asset metadata effectively continues to grow via repeated annotation by the end users. In fact, the viewers become authors in a sense; they alter the content and add value for subsequent viewers. So there is a feedback loop for additional metadata annotation. An interesting example of this phenomenon is dotsub.com where users translate Web media and create subtitles that are made available via a flash player. One may even consider the act of viewing an asset as a source of additional metadata about the asset. Systems can log the number of views, fast forwards, etc. for each asset. Similarly, we may consider systems where the annotators are not casual Web users seeking entertainment, but rather professional annotators logging content, perhaps with specific business purposes in mind. The architecture is somewhat similar, in which an asset is entered directly into the database, and the metadata is added later. For these systems, content management systems (CMS) architectures may be employed where a bus connects various Web service enabled components and workflows are defined for a range of applications (Fig. 4.5.) The terms Digital Asset Management (DAM) or Media Asset Management (MAM) are also used to indicate a more specific type of CMS. Additional automated post processing operations such as importing transcripts or subtitles which may not be available at the time of ingest can be implemented as independent execution threads – reading, processing to create additional metadata, and rewriting to update the asset record. This decoupled processing is used for cases where real-time processing is not practical given existing hardware resources, and where content is not arriving at a continuous rate.

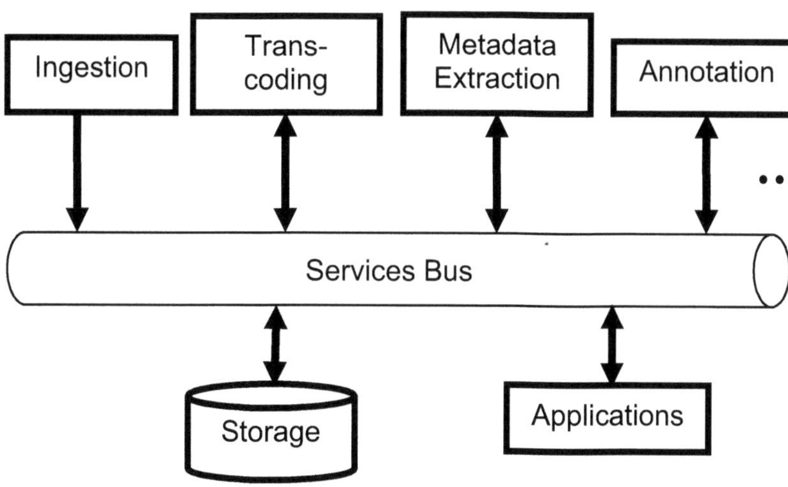

Fig. 4.5. Services oriented architecture for content processing.

The range of options for metadata extraction operations on media is vast and still growing as new media analysis algorithms are developed. These will be discussed in detail in later chapters, but will typically involve demultiplexing streams, decoding and performing computationally intensive operations. Also, with any system that handles video, data bandwidth is a concern and careful system design is required to minimize data access and transport throughout the system. Unfortunately, it is challenging to design a flexible system that can operate at scale to support comprehensive video search. As a result, many search engines today deal solely with high-level metadata, and the extent of their media processing is limited to transcoding and representative image selection.

4.4 Retrieval

The Extensible Markup Language (XML) is well suited for representing the extracted metadata, along with any metadata that accompanies the source asset and any additional metadata that is added over the life of the asset while in the repository. While many metadata systems can be used for high-level metadata, there is really only a single standard intended for representing media features for content based indexing applications, MPEG-7. For audio indexing, [Kim05] describes representing spoken content descriptors in MPEG-7 as well as low-level audio features and their use in classification and similarity metrics.

For efficient handling by the operating system and streaming media delivery systems, several files may be used for each asset – one for the encoded media representation for distribution, a JPEG thumbnail file for browsing and a metadata file in XML format. Of course, other optimizations are possible: for example, a system can be designed to minimize the number of individual files by embedding metadata within the stream. The thumbnail image can even be embedded as a digital item in the stream or key images can be extracted from the video dynamically. The commonly encountered trade off between compute and storage resource applies here as well, and schemes that create and cache temporary files as needed may be efficient solutions. For example, consider a video sharing site that uses Flash as the primary distribution format, but also supports downloading MPEG-4 versions for use with portable media players. The service can transcode all content for rapid response at the expense of storage, or transcode on demand in response to users requests. In the latter case, for popular videos where many user transcode requests are received, transcoding need only be performed once and a cached version of the file will service subsequent requests in order to reduce system compute load.

The XML metadata representation is ideal for transferring information between systems, and for archival storage where additional metadata may be added over the life cycle of the asset. However, for performance reasons, many systems perform a translation from XML into a traditional DBMS (database management systems) approach, at least for high-level metadata. Generating index structures for more fine grained rich media metadata is a research topic in itself and efficient solutions can be achieved for certain cases. Traditional database systems are optimized to respond to queries on multiple fields and return exact matches. A match is well defined, and may be extended to include a range or to support some invariance such as ignoring text case. These systems can be successfully used for tag-based multimedia retrieval, but for content-based retrieval we must extend this capability to support similarity search. Note that we desire semantic similarity which may be subjective and at any rate, can only be approximated algorithmically today. Many multimedia DBMS store features as blobs (binary large objects), perhaps with an application specific similarity metric defined on them. The general problem of constructing indices for rapid retrieval based on high dimensional features is a fertile area of research. For example, [Sant02] explores schema design based on feature substructure to facilitate k-NN and range searches, and Lu [Lu98] discusses performance metrics for multimedia database systems and shows that commonly used metrics may sometimes provide conflicting indications of system performance. For the interested reader, Lu [Lu99] covers many aspects of multimedia database systems design and [Sub98] includes an example of including a movie in a traditional database. Large scale video search implementations can tax even well-designed traditional database systems and lightweight efficient approaches designed to cope with scale [Greer99] can be effective solutions, particularly when extended with support for heterogeneous data represented in XML [Amer02]. It is important to bear in mind that video archiving with mainly textual queries is not a typical DBMS task; large swaths of the traditional database infrastructure, such as ensuring transactional consistency, are not required. For Web search, the act of deleting a record is so infrequent that it is almost not a requirement to implement.

4.5 User Perspectives

4.5.1 Interaction States

Let's consider how a user interacts with a video search engine. We can model this as a set of behavioral states as shown in Fig. 4.6 where the retrieval activities are shown in white and the contribution activities are shown in grey. The states are as follows:

- **Q:** Query – the user is presented with a query interface (e.g. a text box in an HTML form) and must formulate or express the query to the system.
- **B:** Browse – the service presents a set of rank ordered results in response to the query. Metadata and thumbnail images are displayed as a list and the user can interact via scrolling, paging, etc.
- **V:** View – the user has selected a particular item and the system initiates video playback using a media player.
- **A:** Annotation – the user may tag, rate, review or otherwise comment on the video.
- **E:** Edit – the user composes a video using a video editing tool, editing both the content and the metadata.
- **U:** Upload – the user publishes their video content and provides directives for intended audience, content categories, etc.

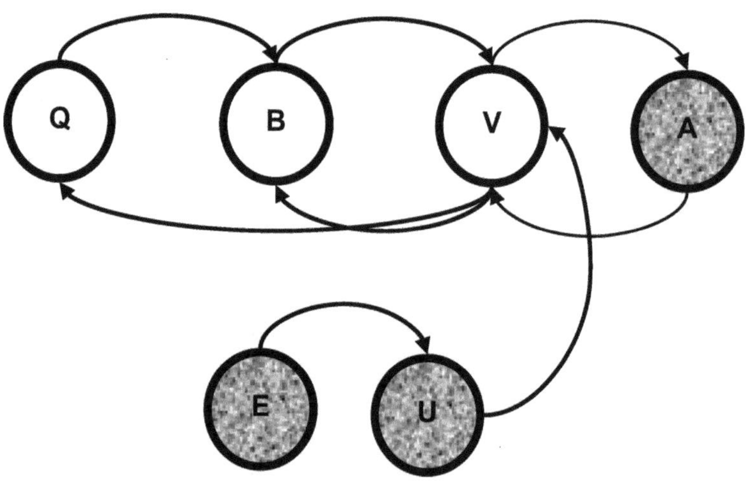

Fig. 4.6. User activities during video search and contribution.

In the figure, only the primary flows are indicated and sites typically allow users to navigate among all the states at will. Also the flow implies a traditional capture–edit–upload content contribution flow (capture is not shown here) which is typical of most user contributed content and professionally produced Web video content. While most video editing today takes place locally prior to contribution, many sites offer video editing of content stored remotely up on the server, so we have included this in the figure where it is implied that there are Web applications that support each state. This user interaction flow follows the classical model (which can be exploited to improve performance [Agi06]) but does not capture concepts commonly employed such as personalization based on user preferences, or the notion of a portal displaying popular or promotional content. The latter case can fit the model here as a special case of the browse state which the user enters with a null query or as an initial state. Further, many systems are constructed to support parallelism in the user interaction. For high-performance retrieval applications, or for immersive entertainment focused applications, perhaps where full-screen video replay is employed to provide a lean-back, TV-like experience, one or more activities can take place simultaneously. In this scenario the video is "always on" and the user may guide the thread of replay using queries or browsing to other selections as an alternative to using the "channel-up" button on their TV remote control.

4.5.2 Granularity of Search Results Representation

As we imagine the user navigating through this list of relevant content identified from a vast sea of video material, we can think in terms of a play list – or an edit decision list. In the former case, the system selects content in response to user queries, and rank orders them for replay to the user. In the latter, relevant segments are identified and selected including "in" and "out" points.

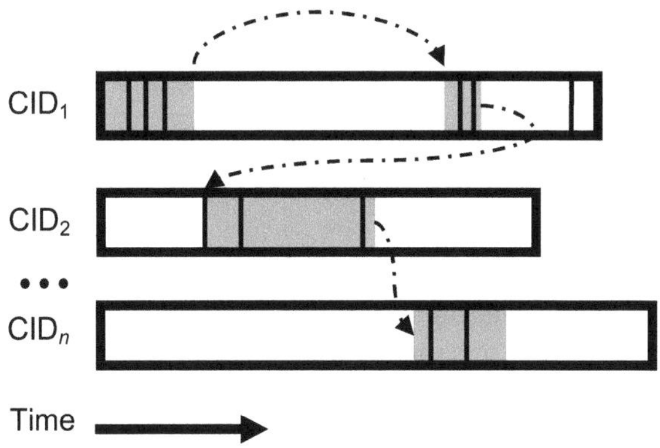

Fig. 4.7. Levels of granularity for representing video search results.

As shown in Fig. 4.7, systems may supply additional information to represent the matches to a user query. The figure depicts a result set with content of different length and a potential path (dashed line) for playing out the media to the user. The levels represented here are:

1. Sets of content identifiers (CIDs) indicating which assets in the database match the user query.
2. Lists of clips specified by offset and duration or in and out points (shown in grey) indicating the most relevant segments of each media file.
3. Lists of "hits" or feature-level matches. For text, using word features, these are the matching words or phrases to the query; in the case of a high-level image concept, these may be matching video frames. So these "hits" while represented in the figure as impulsive events, may indeed have an implicit duration, albeit small – at the word or frame level.

Note that the figure implies a binary (thresholded) decision as to what constitutes a match, but systems may also preserve a measure of the likelihood of match for each level. Application designers must bear in mind that the accuracy of such measures may be difficult to determine accurately. In our example, we may decide that the second clip in CID_1 is of lower rank and should be omitted from the playback. We have lists of identifiers, and lists of temporal intervals with measures of match value. Further, the portion of the query that generated the match can be represented, as can be the portion of the content (e.g. the spatial coordinates of a region of interest of an image containing a face that matches a query.) For practical reasons, many systems discard this detailed query match information as quickly as possible during the stages of query processing and rendering of results.

4.6 Factors Concerning Scalability

4.6.1 Introduction

In the process of designing a content indexing and retrieval system, several factors influence the scalability of the system and basic choices can have a great effect on the cost required to support a given user base. Of course, the fundamental design decision of referring video playback requests back to the originator as opposed to keeping a local transcoded copy of the video results in an entirely different class of service with commensurate costs of operation. The following list of scalability factors assumes a contributed model with local storage, but most of the basic factors relate to a wide range of video search applications:

4.6.2 Acquisition

- **Content arrival rate / variability:** On average, how many assets are posted to the system for indexing for a given time interval? How does this vary over the course of a day or week? What is the peak arrival rate during busy hours?
- **Content average duration:** The duration of the content (posted video clips) effects the processing time.
- **Aggregate incoming content bit rate:** Low bit rate clips will reduce required incoming bandwidth, but the resulting lower quality will negatively effect content based indexing performance and transcoded video quality. Google video suggests using the highest quality available to typical consumers (5M MPEG-2, or 2M MPEG-4) [Goo07].

4.6.3 Processing

- **Real-time factor:** as mentioned above, the types of processing indexing operations can vary widely, from simple metadata ingest to sophisticated feature extraction and content analysis. Transcoding is also considered to be a media processing operation. Once a particular palette of processing operations is selected, there are tradeoffs within each in terms of accuracy vs. speed. A key performance metric is the real-time factor, or ratio of wall clock time to media time; to put it another way, how long does it take to process content of a given duration on a typical server?
- **Latency constraints:** The number of processing servers required is a function of not only real-time factor, content arrival rate and variance, but also the maximum allowable delay from content acquisition to publication. Continuous content acquisition and processing applications such as broadcast monitoring are characterized by a fixed system latency, but for user contributed or syndicated sources, the service can be configured to accommodate the average content arrival rate. Users who post content during busy periods will experience more delay in processing, and the perceived quality of service is a function of the maximum and typical processing delays that a user experiences. Note that for international services, there may be too few appreciable quiet periods for services to exploit to effectively “catch up” (reduce processing queues). Also, processing priority schemes can improve the global user experience, at the expense of a small number of users. For

example, instead of a first come, first served (FIFO) policy, short duration content can be processed before longer content, or new content from a traditionally popular source can be given higher priority.

4.6.4 Storage

- **Archival media bit rate and encoding parameters:** Regarding storage, the single most important factor by far is the size of the bulk media which is governed by the transcoded bit rate, or if the original content is preserved, the policies regarding uploaded media such as allowable bit rate and media duration. Many systems effectively archive the distribution format, that is to say that the distribution format is created once and maintained as the primary source in the archive. In this case, we must take into account considerations such as bit overhead for support of random access (such as short GoPs), forward error correction or bitrate scalability.
- **Alternative media representations:** Are transcoded versions of each asset to be created to support a wide range of devices for playback? For the best user experinces across all devices today, multiple streams are required.
- **Broweable representations:** Key frames used to create visual interfaces for users are the next most significant storage cost. In the extreme, it is possible to omit these altogether, and perhaps maintain an icon image for each series of programs, but this greatly detracts from the overall user experience. Some options for typical solutions include:

  - a single thumbnail for each asset;
  - key frames sampled uniformly or based on the video content;
  - visual summaries that include motion video.

  For each of these options, the spatial resolution of the representative images is an important design tradeoff between storage and quality of user experience. A similar tradeoff exists regarding the number of retained key frames for each asset. Longer video summaries may help users identify relevant videos in query results and additional key frames will increase the precision with which users can visually position long form content for playback. Note that to save long-term storage space, it is possible to dynamically extract these representations from the bulk media as the visual interface is rendered, perhaps with caching but this additional complexity is not typically justified. However, visual browsing within a single asset, using a more capable media player to render visual navigation points is practical.

- **Content description bit rate:** In many existing systems such as RSS aggregation search engines, the number of bytes used in the XML representation of the metadata is negligible in comparison to the media since only global, high level tags are used. However, as more and more timed metadata is included such as chapter titles and transcripts, the size of the description will grow in proportion to the length of the content, and we may speak of a content description bit rate. While textual descriptions compresses effectively using MPEG-7 BiM for example, other, lower level media features such as phonemes or lattices may not compress as readily. It is even possible for the content descriptions to exceed the size of the content (for example, if phonetic lattices are used for searching 8kHz telephone calls.)
- **Index storage:** Derived from the content descriptions whether the form of an inverted text index or other structures for efficient retrieval of binary features, the size of the index eventually becomes an issue as the scale of the content grows. The index is generally maintained in a combination of high performance storage and memory so controlling the index size is a key system performance driver.

4.6.5 Retrieval

- **Peak simultaneous users:** As in any Web application, the primary metric determining scalability of the retreival subsystem is the number of simultaneous users at peak usage times. We may further specify this as the number of users supported per server since replication with load balancing can be used.
- **User activity duty cycle:** The retrieval user interface can be crafted such that the user spends varying amounts of time performing the actions of query, browse, and viewing video (see Fig. 4.6. **User activities during video search and contribution**) for a given session, and these states consume different sets of system resources. For example, viewing video puts no load on the query engine, but taxes the media delivery subsystem. On the other hand, if appropriate context is provided for query results, replay requests for undesired video can be minimized.
- **Output user interface bandwidth:** Rendering rich media user interfaces heavy with still images or flash animations can result in large “pages” or a large amount of interactive content outbound from the service.
- **Output media bandwidth:** By far the largest overall consumer of outbound bandwidth is the replay of the media itself. At its core, video

search is essentially a VoD application and optimizations for system resource utilization (caching of popular content based on Zipf distributions, etc.) have been well studied [Chang97, Yu06]. In addition to the codec selection and transcoded bitrate, other transcoding parameters such as VBR or selection of short GoP to improve random access will effect the video quality vs. output bandwidth.

- **Rate of stream control requests:** Streaming systems supporting rapid start may burst data at a higher rate than the average bit rate at start or after seek requests. If the service is designed such that users rapidly request many videos, as opposed to passively watching long-form video, the load on the server and the total output bandwidth will be increased.

4.7 Retrieval Interfaces

As with text search engine systems, video search systems may support several interfaces to allow developers to create new applications, or for users to access the service using multiple client applications other than browsers such as RSS readers, gadget containers such as desktop sidebars, or mashups which combine multiple services.

Systems may support interfaces including:

- Web Services (WS) – The simple object access protocol (SOAP) used over HTTP and described by the Web Services Description Language (WSDL) provides a standard framework for interaction with services such as video retrieval, but the latitude enabled in architecture and parameter semantics implies that video search Web services are not interoperable in general. Some amount of new interface code is typically required when switching from one Web service to another.
- Representational State Exchange (REST) – The REST presents architectural style [Field00] that can be applied to media search engine interactions. Flickr® for example, supports a REST API for photo retrieval. This design philosophy is highly robust (it describes the behavior of stateless exchange on the Web for example) but does not dictate a preferred language or syntax.
- Asynchronous JavaScript with XML (AJAX) – AJAX can be used for rendering dynamic user interfaces, and the underlying requests to the video Web server may have a REST flavor and may utilize WS.
- Real Simple Syndication (RSS) – Query results can be represented in XML using RSS, perhaps with specific namespace extensions as appropriate. This enables client-side persistent queries, in which the

results are typically sorted temporally (most recent first) to best fit the usage paradigm of RSS readers.

- OpenSearch® – A9, an amazon.com company, developed OpenSearch® which provides a standardized query syntax that allows browsers and other applications to seamlessly support a wide range of search engine providers [Clin07]. Results are formatted in RSS and a paging mechanism is supported to allow stepping through large results lists efficiently.
- SQL / XML Query – A video collection may expose an SQL interface and applications may connect using ODBC. The Microsoft Indexing Service is an example of a service that supports an SQL interface but with an underlying architecture that is more indexing oriented than a traditional database. Several XML query languages are available for use with metadata stored in XML such as MPEG-7.
- Mini-applications (Gadgets, Widgets) – Microsoft's Vista®, Google and Yahoo support interfaces to allow third parties such as video search engines to expose functionality through a small region of screen real estate in other applications or desktops. These are typically grouped via a sidebar.
- Notification, e-Mail – Systems can support persistent queries stored on the server and notify users via e-mail. Users can elect to be notified once a day rather than multiple times as new matching content arrives. Servers must efficiently manage stored queries from a potentially large set of users, and possibly pool similar queries.

4.8 Typical System Features

We indicated notification as an interface, but this may be thought of as more of a "service" or feature than an API. Other features typically found in video search services include:

- **Favorites** – saved searches where results may appear on the search engine landing page freshly executed with each visit.
- **History** – access to videos recently viewed.
- **User preferences** – capture and support user preferences such as explicit content filtering. In some cases bandwidth or media player preferences are stored as a cookie.
- **Cut-and-paste links** – The service may make available markup suitable for rapid cut-and-paste into blogs or other applications to allow either

link back to the video search engine with the video of interest selected, or to insert a video plug-in to deliver the video.

- **Download** – The ability for users to download the video with transcoding as appropriate for particular devices.
- **Annotation** – ability for users to post persistent comments, rate or tag videos and make these annotations searchable.

4.9 Conclusion

We've seen that the basic structure of video search involves acquisition, media processing and retrieval systems. This basic architecture has parallels in video on demand systems and media asset management (MAM) systems, albeit with different optimizations for the scale and features desired. We presented options for video search engine architectures and supported features and APIs along with the associated tradeoffs for system resource utilization. Architectures have evolved as new technologies have become practical and widespread and we will no doubt continue to see advances in video search architectures going forward.

References

[Agi06] Agichtein, E., Brill, E., and Dumais, S.: *Improving Web search ranking by incorporating user behavior information,* in Proceedings of the 29th annual international ACM SIGIR conference on Research and development in information retrieval. New York, NY, USA: ACM Press, pp. 19–26 (2006).

[Amer02] Amer-Yahia, S., Fernández, M., Greer, R., and Srivastava, D. 2002. Logical and physical support for heterogeneous data. In *Proceedings of the Eleventh international Conference on information and Knowledge Management,* CIKM '02. ACM, New York, NY, pp. 270–281 (2002).

[Bec04] Becarevic, D., and Roantree, M.: A metadata approach to multimedia database federations, Information and Software Technology, **46**(3), pp. 195–207 (2004).

[Cha03] Chakrabarti, S.: *Mining the Web,* Morgan Kaufmann Publishers. (2003).

[Chang97] Chang, S.F.: *Video on Demand Systems Technology, Interoperability, and Trials.* Boston: Kluwer Academic Publishers (1997).

[Ciln07] Clinton, D.: Specifications/OpenSearch/1.1/Draft 3 – OpenSearch, http://www.opensearch.org/Specifications/OpenSearch/1.1, cited 12 Feb 2008.

[Field00] Fielding, R.: *Architectural Styles and the Design of Network-based Software Architectures.* Doctoral dissertation, University of California, Irvine (2000).

[Goo07] Google, What you're your technical requirements and quality guidelines for uploaded videos? Google video help center, URL: http://video.google.com/support/bin/answer.py?answer=26562, cited 16 Jun 2008.

[Greer99] Greer, R.: Daytona and the fourth-generation language Cymbal. In *Proceedings of the 1999 ACM SIGMOD international Conference on Management of Data* SIGMOD '99. ACM, New York, NY, 525–526 (1999).

[IBM07] IBM Smart Surveillance System (Previous PeopleVision Project) http://www.research.ibm.com/peoplevision/index.html, cited 21 Dec. 2007.

[Keht06] Kehtarnavaz, N., and Gamadia, M.: *Real-Time Image and Video Processing From Research to Reality.* Synthesis lectures on image, video and multimedia processing, #5. San Rafael, Calif. Morgan & Claypool Publishers (2006).

[Kim05] Kim, H.-G., Moreau, N. and Sikora, T.: *MPEG-7 Audio and Beyond Audio Content Indexing and Retrieval.* J. Wiley, Chichester, West Sussex, England (2005).

[Kosch03] Kosch, H.: *Distributed Multimedia Database Technologies Supported by MPEG-7 and MPEG-21*, CRC Press, Boca Raton, Florida (2003).

[Lu98] Lu, G. and Sajjanhar, A.: On performance measurement of multimedia information retrieval systems, International Conference on Computational Intelligence and Multimedia Applications, Monash University, Gippsland Campus, pp. 781–787 (1998).

[Lu99] Lu, G.: *Multimedia Database Management Systems*, Artech House Publishers (1999).

[Sant02] Santini, S., and Gupta, A.: Principles of Schema Design for Multimedia Databases, *IEEE Transactions on Multimedia*, **4**(2) (2002).

[Sub98] Subrahmanian, V. S., and Tripathi, S. K.: *Multimedia Tools and Applications. Vol. 7, Nos. 1/2, Multimedia Information Systems.* Boston: Kluwer (1998).

[Yu06] Yu, H., Zheng, D., Zhao, B. Y., and Zheng, W.: Understanding user behavior in large-scale video-on-demand systems. In *Proceedings of the ACM Sigops/Eurosys European Conference on Computer Systems,* pp. 333–344 (2006).

5 Media Processing

5.1 Introduction

The only media descriptions available for the vast majority of media published on the Web today is global high level metadata. To differentiate themselves from systems that treat the described media payload as an opaque data file, individual systems must employ automated content processing. While specific processing methods have been optimized to handle particular media types, there are common principles that apply to some degree across all media types. The field of digital signal processing includes several areas of focus including speech, audio, image, and video processing. If we stretch the notion of signal processing from digitizing an analog waveform to include streams of symbols, we can consider text streams corresponding to the media dialog to be signals as well [Rab99]. Common media processing operations include noise reduction, re-sampling, compression, segmentation, feature extraction, modeling, statistical methods, summarization, and building compact representations for indexing and browsing.

In the previous chapters we discussed the practical issues of compression systems in use today as well as container file formats for media streams. We discussed media related text streams and formats including closed caption, subtitles, transcripts, etc. Here we will present at an introductory level, the common elements for media processing as it relates to content-based video search engine systems. In later chapters, we will explore in greater detail some of the most common methods applied to audio, video and text streams, and we will present multimodal processing where these media streams may be processed in a coordinated manner to achieve greater accuracy than is possible by processing the components individually.

As we look into each media type in more detail, we will focus on feature extraction, segmentation, and information extraction. For us, the desired goal of media processing is to take largely unknown content and extract some level of structure and possibly semantics about the content. In the case of data mining, we might hope to obtain actionable information based on this analysis. We will find that low-level feature extraction has been well studied and robust, efficient methods exist for operating on multiple media types, but moving to true semantics or meaning is successful only in restricted domains – where we have some domain-specific knowledge and perhaps have developed models based on similar labeled data. This difficulty with the current state of the art in moving from low-level features to useful understanding of the media content is often referred to as the semantic gap.

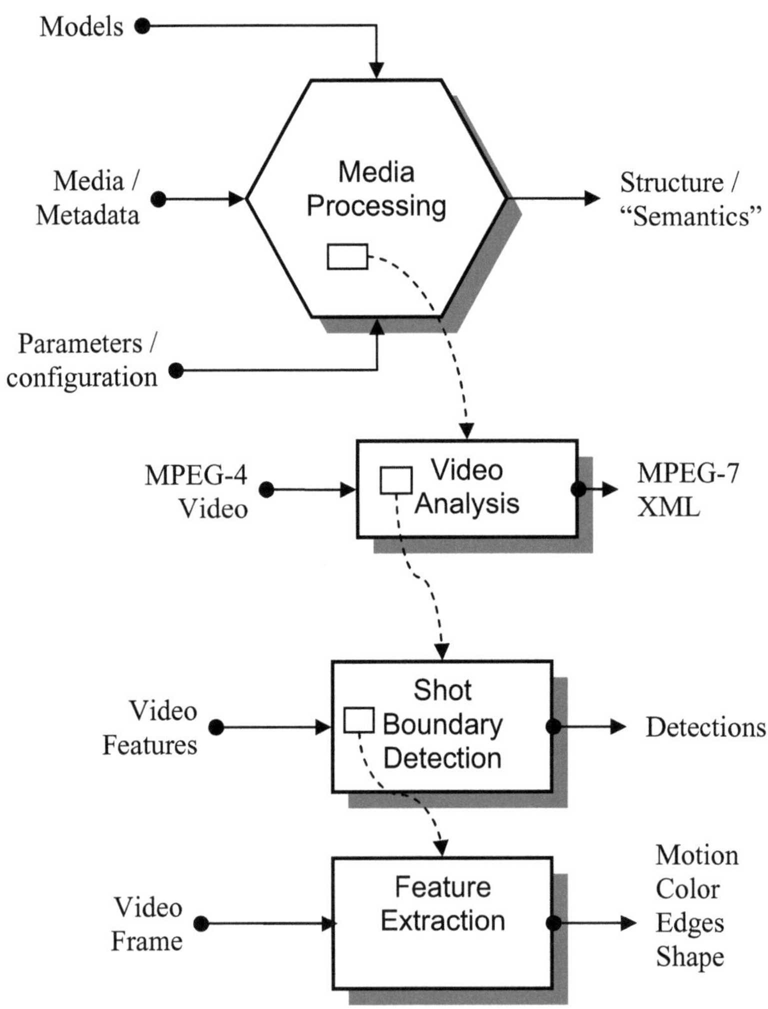

Fig. 5.1. Conceptual view of media processing.

Fig. 5.1 represents a hierarchal view of media processing for video retrieval applications and focuses on the case of shot boundary detection in particular. Each level is characterized by functional blocks with representative input and output data types shown. The scope of media processing is broad indeed, when one considers that similar "drilldown" views could be drawn for each of the other tasks such as speaker identification, text-based topic segmentation, etc. Representative features of color, shape descriptors, etc. are shown and MPEG-4 is shown as an illustrative media source format. The results may be represented in an XML format such as MPEG-7 as the figure suggests.

5.2 Feature Extraction

Feature extraction implies processing series of media samples (signals) and creating more efficient representations that will be eventually useful in deriving meaning about the content represented by the signals. Generally for media processing, this achieves a large data reduction, sometimes on the scale of several orders of magnitude. It is typically not practical or useful for all stages of media processing to operate directly on the samples that are intended primarily for regenerating the signal. The extracted features can be straight statistical measures such as mean or moments, but in many cases the features are intended to model human perception or physiology so more advanced transforms are used (e.g. Mel Frequency Cepstral Coefficients). Note that the task of reducing data while retaining perceptually meaningful information is not only the goal of this stage of media processing, but is also the goal of media compression. Therefore, in many cases, the same features are used and the theoretical basis as well as the algorithms and implementations can be re-used. In fact, many practical media processing algorithms are designed to operate in the compressed domain. While the features optimized for compression are not necessarily optimal for analysis, they are reasonably good and a large measure of systems efficiency can be realized by their adoption to provide double duty in analysis as well as compression. Although we mentioned the importance of data reduction, often the first operation for generating features can be a transformation, e.g. from the time / spatial domain to the frequency domain which is not necessarily inherently lossy. It is just that in this new space it may become more straightforward to truncate features in such a way that, when the inverse transform is preformed, results in minimal perceived signal degradation. Thus we can speak of a feature space in which samples are represented as feature vectors. Typically we would like to keep the dimensionally of the vectors as small as possible to improve system efficiency and perhaps generalizability, but in some cases high dimensionality is not an insurmountable problem, particularly when the feature vectors are sparse. One notable exception to the notion of data reduction is the case of query expansion, where there is typically a paucity of features – the source data may be a single word entered by a user – so it is desirable to appeal to ancillary data sources in an attempt to create additional features that may capture the intent of the query.

The problem of feature selection arises simply because it is generally easier to generate features than to determine which features have value for a desired application. It may not be computationally practical to use all extracted features, and in fact using all features may have a detrimental effect on the accuracy of the results.

In algorithm design, the invariance of the features must be taken into account. In image processing, scale, rotation and translation invariance are generally desirable. However, there may be limits imposed. For example, we may build a face detection system based on spatial relations of low level features, and then use the presence, location and orientation of the detected faces as higher level features for later processing in story segmentation. We desire our face detector to be invariant to slight rotations of the face about an axis perpendicular to the image plane, but not if that rotation is on the order of 180 degrees (where the detector reports that the face is upside down – this is either an error or a situation of no relevance).

5.3 Media Segmentation

Segmentation is important for a number of reasons. The concept of segmentation is to divide a stream of media into semantically consistent units and one benefit is increased efficiency in representation or compression. For example, in video compression shot boundary detection can be used so that difference frames are calculated within a shot instead of across shot boundaries, resulting in much smaller deltas. For representing content to users for rapid browsing, it is often desirable to remove the temporal element, i.e. represent long, static segments with a single icon, but also include shorter segments with similar icons, such as in a light-table view. While the temporal aspect is not preserved, the viewer can immediately conceive of the basic semantic content of the video without having to parse redundant information. Segmentation also benefits information retrieval since metrics such as TF/IDF are more accurate after text has been segmented by topic. Consider a news program with five stories, one of which mentions NASA and the space shuttle many times, while the other stories are unrelated. The calculated relevance rank of the program will be reduced if the frequencies of occurrence are averaged over the entire program.

One of the challenges for practitioners of media processing in the context of segmentation is to determine the appropriate level of granularity, or if multiple levels are to be maintained in the system, what is the appropriate number of tiers in the hierarchy. At the base of this pyramid, we have the media samples themselves, and compression algorithms attempt to remove redundancy at this level. Moving higher we can extract low-level features based on small windows of time or space, typically on the order of 10 milliseconds for audio, or corresponding to small homogenous regions of an image. Of course the definition of homogenous is a bit problematic: do we mean the same amplitude, same gradient, or for textures, the same periodic pattern? Moving a bit higher, we enter a realm where the extracted symbols may convey meaning, e.g. phonemes or words from speech recognition systems. Continuing in the speech domain for a moment, we encounter phrase or sentence segmentation tasks, and later topic or story segmentation. For image and video processing, we have object or foreground / background segmentation within a frame, camera operation detection, shot segmentation followed by scene segmentation – where multiple shots may take place in a single physical location. Farther up, for produced video programs, we have program segment (or commercial) detection and again story or topic segmentation, perhaps using cues from multiple media streams. Now, it is implicit that we can stop segmentation when we have reached the top: a single media asset or file. But what about episodic content? Does not each asset instance represent a segment of a longer narrative story where familiar characters reappear and evolve? And are not the productions of similar genre, or from the same source, somehow related? This latter level of segmentation moves us out of the signal processing and statistical classification domain into database organization – typically we have labeled data in the form of an EPG to guide us here.

5.4 Clustering, Structure Generation

Assuming we have good segmentation, the notion of clustering or forming relations between segments using distance metrics arises. Consider a video program where two participants are discussing a series of issues, and we have three cameras in the studio, one for close-ups of each speaker and one for a wide shot. We now successfully detect the cuts between the shots to segment the media into logically consistent chunks at a temporal level on the order of tens of seconds. However it is clear that another level of structure can be derived by analyzing the segmented media in order to associate related segments. In this case, we may discover a pattern that the producer has used to move between cameras: A,B,C,A,B,A,C, etc. where A and B represent the close-up views of each speaker and C is the wide shot. As a second example, consider the case where an editor is trying to produce a rough cut from rushes or repeated 'takes' of a particular scene. We may detect the start and stop of the camera, but we can also discover that there were five attempts to capture the first scene, and then eight attempts of a subsequent scene, etc. By analyzing the relative "distance" (or similarity) between subsequent shots we can derive this structure. In fact commercially available editing systems can perform this function using audio cross-correlation assuming there is repeated dialog in each shot. This can be of great value for navigating and organizing the mass of raw footage during the editing process. This level of organization (aggregating repeated takes of a particular scene with or without dialog) has been the subject of a research evaluation undertaken by the National Institute of Standards and Technologies in a rushes summarization task [Over07]. Summarization via automated content analysis allows users to more easily browse long-form content by removing redundancy within an asset, and clustering across search results sets facilitates browsing of large media archives.

Other important considerations should be borne in mind when evaluating media processing algorithms. Is the method rule-based, data-driven, or some combination thereof? Does the method involve the use of tunable parameters? How generalizable is the method? What are the storage vs. computational performance tradeoffs? For natural language applications, rule-based systems are generally quite useful when training data is not available but may become unmanageable as the complexity increases. Data-driven methods may offer the promise of managing this complexity in a scalable manner, but inevitably suffer from the problems that arise from the mismatch of training data vs. the data encountered in the field. We can expect performance to degrade over time as this gap between training and testing datasets widens. Steps must be taken to adapt the existing models over time or to the new domains. Active learning may be effective to minimize the labeling effort while maximizing performance improvement. For data driven methods the definition of the labels (typically as defined in an annotation guide) and their successful application by the labelers becomes an important factor in system performance. It is generally observed that more labeled data is better, but high quality labeling of large datasets is costly. Again the more-is-better camp will argue that we can ask many labelers to label the same data and use techniques to derive a consensus labeling. Recently it has been observed that the human power of the Web can be exploited, perhaps via game play scenarios, to build up large labeled data collections [Ahn06]. For these un-trained labeler situations, special effort is given to avoid tag synonyms, redundant or inconsistent labels, etc. Also, as we start to tap into the social capabilities available via the Web, we must also consider the practical limits to automated content processing. At some point, if our goals of media understanding are impractical, and if the value of the content is high enough, we run up against the alternative which is manual content description. For example, if a speech recognition system only performs acceptably for broadcast news content, then this system is of little value since most of this material is closed captioned or transcribed already. Many DVD subtitles are translated manually (and voluntarily) and posted up to Websites, rendering the use of machine translation systems in this domain moot.

5.5 Real-Time Processing

For real-time processing applications, resource management is critical in order to achieve optimal performance. Distributed systems may be employed, but careful consideration of data bandwidth utilization is paramount. Event-based architectures can provide efficient system design and fit in well with publish/subscribe models [Wold02]. For user-contributed applications, we don't necessarily have the issues associated with real-time processing of incoming media streams, but we have a larger scale issue of operating on received content in a timely manner to provide desired results to the user. In this case, systems can be designed to produce partial results to provide rapid user feedback while more time consuming operations can be deferred, perhaps to off-peak hours.

5.6 Systems Issues and Architectures

The media processing systems programmer can benefit from using one of the several frameworks that have been developed primarily for rendering and transcoding media of various formats on different platforms. Good modular systems design has led to implementation of dataflow architectures employing sources, sinks, and filters. These frameworks include the open source project gstreamer [Kata06], [Taym07], Microsoft's Direct Show [Pesce03], and Apple's QuickTime [Hoff92]. Third party vendors may supply their own filters for decoding a particular media format, parsing a container format, or demultiplexing audio and video streams. Common rendering and transcoding operations such as cropping and scaling are implemented as pass-though filters that operate on a particular media type. The architecture does not impose restrictions on filters (such as forbidding file I/O for parameter specification or results storage), and the filters may utilize hardware acceleration when available. Media processing including elemental feature extraction all the way up through higher-level functionality such as face detection or shot boundary detection can be implemented as filters. This framework relieves the burden on the programmer of supporting the myriad of media formats. These systems can operate in real-time where a clock is derived from a capture or rendering filter or in a free-running mode for faster than real-time operation. Systems include the notion of input/output pins or pads where their datatype must match in order for a connection to be made (e.g. uncompressed RGB pixel values) and may include some ability to automatically invoke and connect suitable transform filters if the pins are of different types. A typical example of this is a decoder that outputs Y,Cr,Cb and a rendering device that only supports RGB: a color space conversion filter may be added automatically to make the necessary transformation.

5.7 Conclusion

We provided a general introduction to media processing for video retrieval applications. We looked at common aspects for audio, video, and text media processing algorithms including computing features to represent the media and facilitate later processing, segmentation into logically meaningful units at different temporal levels, and touched on distance metrics, clustering and browsing. These topics will be explored further in the chapters that follow, which are organized with respect to media type.

References

[Ahn06] Ahn, L.: Games with a Purpose. *Computer* **39**(6), pp. 92–94 (2006).

[Ebroul06] Ebroul, I.: Editorial: Knowledge-based digital media processing, *Vision, Image and Signal Processing, IEEE Proceedings*, **153**(3), pp: 253–254 (2006).

[Hoff92] Hoffert, E. et al.: QuickTime: an extensible standard for digital multimedia *Compcon Spring '92. Thirty-Seventh IEEE Computer Society International Conference, Digest of Papers*, Apple Inc., Cupertino, CA (1992).

[Kata06] Katafiasz, M.: Multipurpose multimedia processing with GStreamer – A universal solution for your multiple needs, http://www-128.ibm.com/developerworks/aix/library/au-gstreamer.html?ca=dgr-lnxw07GStreamer (2006).

[Krik97] Krikelis, A.: Multimedia processing architectures, *IEEE Concurrency*, **5**(3), pp. 5–7 (1997).

[Over07] Over, P., Smeaton, A. F., and Kelly, P.: The trecvid 2007 BBC rushes summarization evaluation pilot. In *Proceedings of the international Workshop on TRECVID Video Summarization,* ACM (2007).

[Pesce03] Pesce, M.: *Programming Microsoft DirectShow for digital video and television*, Microsoft Press (2003).

[Rab99] Rabiner, L., personal conversations (1999).

[Steinm02] Steinmetz, R., and Nahrstedt, K.: *Multimedia Fundamentals, Volume 1: Media Coding and Content Processing*, 2nd ed, Prentice Hall (2002).

[Taym07] Taymans, W. et al.: GStreamer Application Development Manual, 0.10.14.1 (2007).

[Wold02] Wold-Eide, V. et al.: Real-Time Processing of Media Streams: A Case for Event-Based Interaction, *22nd International Conference on Distributed Computing Systems Workshops* (ICDCSW '02) (2002).

6 Video Processing

6.1 Introduction

Along with the expeditious improvements in the network, computation, storage technologies and consumer electronics, video content has become much more commonplace, and more accessible. Powered by the broadband wireless connection capabilities provided by Wi-Fi, 3G, WiMax, etc., and the video playback capabilities available on various personal digital assistants (PDA), the vision of searching and browsing video content anywhere and anytime is becoming a reality.

Video content indexing and retrieval has been an active research area for the last two decades [Dimitrova02]. Although it is still far from mature, many practical multimedia query systems have been successfully built. Representative image and video search systems include QBIC [Flickner95] and CueVideo [Ponceleon98] developed by IBM, CuVid [Chang07] built by Columbia University, VideoLogger [Virage07] provided by Virage, and the Miracle system [Gibbon06] developed by AT&T Labs. The rapid advance in video content analysis and delivering technologies allows the service providers to offer more personalized and interactive video services to the end users, for example, IPTV services [Xiao07].

Recently, Web search and online advertising service companies, including Yahoo, Google, and MSN have begun to provide online search services for multimedia content, including audio, image, and video. The most successful video sharing website is YouTube [YouTube07], which attracted millions of users around the world. Flickr [Flickr07] is one of the most popular websites for sharing and managing photos. Although most of these search engines are purely relying on metadata attached to the content or annotation created by the users, they have been proven to be very useful.

The challenge of locating relevant content in a sea of multimedia documents demands a standard solution to identifying and managing content. MPEG-7 offers a comprehensive set of audiovisual description tools to create descriptions, which will form the basis for applications enabling the needed effective and efficient access to multimedia content [Martinez02a, Martinez02b]. Content described by MPEG-7 will be inherently more valuable because it will be more easily searchable and accessible.

TRECVID [Kraaij06], sponsored by NIST, further stimulates the interest and effort in automatic segmentation, indexing, and content-based retrieval of digital video in a broad research community. New systems and algorithms have been constantly reported from all TRECVID participants over the years, e.g. IBM, Tsinghua University, Columbia University, CMU, KDDI, etc. So far, TRECVID has organized many evaluation tasks, including shot boundary detection, high-level and low-level feature extraction, story segmentation, search, rushes summarization, content-based copy detection, surveillance event detection, etc. These tasks cover a wide spectrum in the video analysis and retrieval area.

The structure of this chapter is as follows. Section 6.2 focuses on shot boundary detection, which is normally the first component in most video processing systems. We discuss how to select representative keyframes for each shot in Sect. 6.3. Face detection and recognition are described in Sects. 6.4 and 6.5. We briefly introduce the video OCR techniques in Sect. 6.6. Concept detection in video is then presented in Sect. 6.7. In Sect. 6.8, user interface issues, including video browsing and navigation are considered. Finally, we draw conclusions in Sect. 6.9.

6.2 Shot Boundary Determination

Shot boundary determination (SBD) has been widely studied for the last decade. Some of the early work can be found in [Shahraray95, Wang00, Yeo95, Zhang93]. Researchers at AT&T started to tackle multimedia content processing and indexing in the early 1990s, and Shahraray reported a scene change detection algorithm in 1995 [Shahraray95]. With the limited computation power (90M CPU) and system memory (8 MB) available at that time, as well as the constraints of real time and low latency, the original algorithm was designed to be effective and highly efficient. The adopted visual features were intensity histograms and image matching with one-dimensional motion compensation by projection. A single finite state machine (FSM) was designed to detect all types of scene changes and report camera motions, including panning and tilting. An improved version of this algorithm is adopted in the MIRACLE system, a video search engine, at AT&T [Gibbon06].

In this section, we describe the AT&T SBD system [Liu06] evaluated in TRECVID 2006. Thanks to current computational power, there is a lot of room to extend the existing algorithm. Three major improvements are: (1) two-dimensional motion compensation; (2) utilizing color information in addition to intensity values; (3) instead of using a single FSM, multiple FSM-based detectors are adopted to track different types of shot boundaries, e.g. cut, fade in/out, dissolve, wipe, etc. The new architecture is more flexible and modularized: each detector is independently designed and adjusted, and additional detectors can be easily plugged in to capture any new types of shot boundaries.

There are three main components in the AT&T SBD system: visual feature extraction, shot boundary detectors, and result fusion. Figure 6.1 shows the high level diagram of the SBD system. The top level of the algorithm runs in a loop, and every loop processes one video frame. Each new frame and the associated visual features are saved in circular buffers. The loop continues until all frames in the source video, e.g. an MPEG file, are processed.

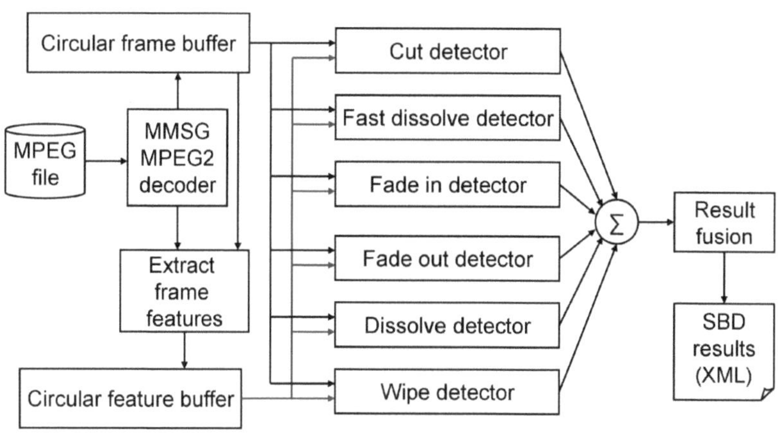

Fig. 6.1. Overview of AT&T shot boundary determination system.

Given the wide varieties of shot transitions, it is difficult to handle all of them using a single detector. The system adopts a "divide and conquer" strategy. Six independent detectors were devised, targeting for six dominant types of shot boundaries in the SBD task. They are cut, fade in, fade out, fast dissolve (less than five frames), dissolve, and wipe. Essentially, each detector is a finite state machine, which may have a different number of states. Finally, the results of all detectors are fused and the overall SBD result is generated in the necessary format.

6.2.1 Feature Extraction

For each frame, the system extracts a set of visual features, which can be classified into two types: intra-frame and inter-frame visual features. The intra-frame features are extracted from a single, specific frame, and they include color histogram, edge, and related statistical features. The inter-frame features rely on the current frame and one previous frame, and they capture the motion compensated intensity matching errors and histogram changes.

Figure 6.2 illustrates how these visual features are computed. While any resolution source material is supported, the resolution of the TRECVID evaluation sequences is 352×240 pixels. The visual features are extracted from a central portion of the picture, which are called the region of interest (ROI). The ROI is marked by a dashed rectangle in Fig. 6.2, overlaid on the original image. The choice of the ROI size is based on two considerations: (1) The ROI covers the majority of the image and effectively eliminates the letterbox for wide screen content. (2) The ROI avoids the border effect in the following feature extraction steps.

Fig. 6.2. Visual feature extraction for SBD.

Within the ROI, the system extracts the histogram of red, green, blue, and intensity channels and computes a set of common statistics, including the mean, the variance, the skewness (the third order moment), and the flatness (the fourth order moment). A visual feature called histogram dynamic range is also extracted, which roughly measures how wide the histogram spreads. To compute the intensity dynamic range, the histogram was searched from both ends, until the accumulated mass of both sides is more than 2%. The dynamic range is the difference of these two values.

For each pixel in the ROI, its discontinuities in the horizontal (with respect to vertical) direction are computed by Sobel operators [Gonzalez93]. If the value is higher than a preset threshold, the pixel is labeled as a horizontal (respectively, vertical) edge pixel. Finally, the ratio of the total number of horizontal (respectively, vertical) edge pixels to the size of ROI is used as an edge based feature.

The temporal derivative (delta) of a feature (e.g. histogram mean) is fitted by a second-order polynomial to make it smooth. The delta values of histogram mean, variance, and dynamic range are computed as additional visual features.

Motion features are extracted based on smaller blocks within the ROI. Specifically, in Fig. 6.2, the ROI (288×192 pixels) is split into 24 blocks (6 by 4), each with the size 48×48 pixels. Generally, motion information extracted from bigger block sizes (e.g. 48×48) is more reliable than those from smaller sizes (e.g. 8×8). The search range of motion vector for each block is set to 32×32. It could be either an exhaustive search for better accuracy or a hierarchical search for higher efficiency. The motion features for each block, e.g. block k, include the motion vector (MV_k), the best matching error (ME_k), and the matching ratio (MR_k). The matching ratio is the ratio of the best matching error with the average matching error within the searching range, and it measures how good the matching is. The value is low when the best matching error is small and the block has significant texture. Based on the motion features of all blocks, the dominant motion vector and its percentage (the ratio of the number of blocks with this motion vector to the total number of blocks) are used as frame level features. The system then ranks all ME_k (resp. MR_k), and computes the order statistics, including the mean, ME_A; the median, ME_M; the average of the top 1/3, ME_H; and the average of the bottom 1/3, ME_L (resp. MR_A, MR_M, MR_H, MR_L). These features are effective in differentiating the localized visual changes (e.g. foreground changes only) from the frame-wise visual changes. For example, high MR_H with low MR_A indicates a localized transition.

In total, 88 visual features are extracted for each frame. Interested readers can find more details in [Liu06].

6.2.2 Shot Boundary Detectors

Figure 6.3 illustrates the general FSM structure for all shot boundary detectors. State 0 is the initial state. When the transition start event is detected, the detector enters the sub FSM, which detects the target transition pattern, and locates the boundaries of the candidate transition. If the sub FSM fails to detect any candidate transition, it returns to state 0, otherwise, it enters state N. State N further verifies the candidate transition with more strict criteria, and if the verification succeeds, it transfers to state 1, which indicates that a transition is successfully detected, otherwise, it returns to the initial state. Although the six detectors share the same general FSM structure, their intrinsic logic and complexity is quite different. In the rest of this section, we briefly discuss all the individual detectors. For more details, please refer to [Liu06].

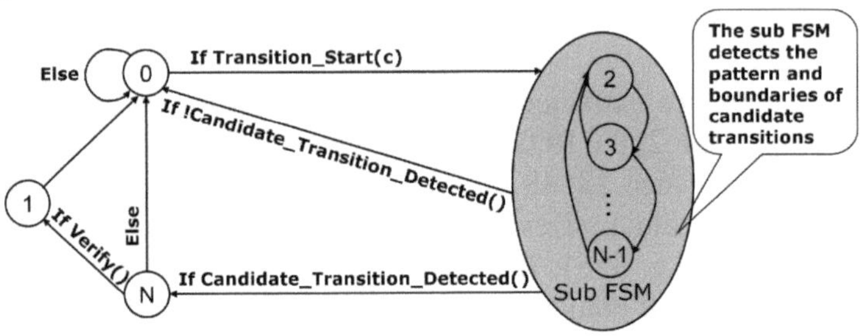

State	Description
0	Initial state
1	Transition detected state
2 ~ N-1	Candidate transition detection states
N	Verification state

State Variables
Current state
Transition start frame
Transition end frame
Detector specific variables

Fig. 6.3. General FSM for transition detectors.

Cut Detector

The cut detector uses a state variable, *AverageME*, to track the average value of matching errors. Its initial value is set to 5.0, and it is updated whenever the state is 0 with the following infinite impulse response (IIR) filter,

$$AverageME = AverageME \times 0.85 + ME_A \times 0.15, \tag{6.1}$$

If the current mean matching error, ME_A, is larger than 5 times *AverageME*, the sub-FSM is activated. The main roles of the sub-FSM are to check whether the candidate boundary has the local maximum matching error, and to introduce a three-frame delay for verification. A verification function, Verify(), compares all pairs of frames in the neighborhood (within three frames) of the boundary, such that false cuts introduced by camera flashes can be effectively removed.

The system also contains a cut verification engine based on a support vector machines (SVM) [Vapnik98]. Assuming k is the end frame of a candidate cut, we extract four groups of features. The first group is the original visual features (88 dimensions) of frame k. The second group is the mean and the standard deviation of all features within an 11-frame window centered at *k*. The third and the last group of features are the same statistics on a 21-frame window and a 31-frame window. All these features are concatenated into a 616-dimensional feature vector as SVM input.

Fade in Detector

Fade in can be reliably detected using the intensity histogram variance. Low variance (not necessarily low intensity) is a strong indicator for the beginning of fade in. Fade in transitions often start from a group of low variance frames and then the variance gradually increases until it becomes stabilized.

The verification algorithm pinpoints the starting and the ending frames of the candidate transition based on the variance value, and it then measures the linearity of the standard deviation (STD) of the intensity. The detector uses *r2* as a measure of linearity in linear regression. Assume there are a set of pairs: $\{x_i, y_i\}$, $1 \leq i \leq N$. By minimum square error, it is straightforward to compute the optimal *a* and *b*, which minimize the error E_{reg}:

$$E_{reg} = \sum_{i=1}^{N} (y_i - ax_i - b)^2 \tag{6.2}$$

and *r2* is defined by

$$r2 = 1 - \frac{E_{reg}}{E_{tot}}, \text{ where } \quad E_{tot} = \sum_{i=1}^{N} \left(y_i - \overline{y}\right)^2 \tag{6.3}$$

If the linearity of the STD curve is higher than a preset threshold, the Verify() function returns true, otherwise, it returns false.

Fade out Detector

Similar to the fade in detector, the fade out detector is also triggered by low variance frames. The verification algorithm checks the linearity of the standard deviation of the intensity. Very often, fade out and fade in transitions are adjacent, and the overlapped fade out and fade in transitions are merged into a single fade out/in (FOI) transition in the result fusion step.

Fast Dissolve Detector

Fast dissolve is triggered by a medium change of the matching error, where ME_A is bigger than $2 \times AverageME$. Let *X*, *Y*, and *Z* denote the starting frame, the ending frame, and a middle frame within a fast dissolve transition. It is required that the duration of the fast dissolve transition be less than five frames, so it is reasonable to assume that there is no motion involved in the transition. With this assumption, Z can be written as a linear combination of X and Y, $Z = \alpha X + (1 - \alpha)Y$, where $0 \leq \alpha \leq 1$. The value of α can be determined by a minimum square error criterion. If the fitting error is smaller than a preset threshold and $0.2 \leq \alpha \leq 0.8$ for all middle frames of the transition, then the Verify() function returns true.

Dissolve Detector

A dissolve is a procedure of linearly mixing two different scenes X and Y. Assuming Z_i is an intermediate frame, then we can use the following formula to represent Z_i,

$$Z_i = \alpha_i X + (1 - \alpha_i)Y \tag{6.4}$$

where $\{\alpha_i\}$ are a set of monotonically increasing values that are in the range [0, 1]. Let the variances of X, Y, and Z_i be σ^2_X, σ^2_Y, and σ^2_{Zi}. If we also assume X and Y are independent, then we have

$$\sigma^2_{Z_i} = \alpha_i^2 \sigma_X^2 + (1 - \alpha_i)^2 \sigma_Y^2 \tag{6.5}$$

If $\sigma^2_X = \sigma^2_Y$, the curve for σ^2_{Zi} is a symmetric quadratic function, shown as in Fig. 6.4(a). But in typical cases, the curve is more like that shown in Fig. 6.4(b), where σ^2_X is not equal to σ^2_Y, and X and Y are not independent. When the variance of either X or Y is small, the variance curve may only contain either the decreasing or the increasing pattern, such as illustrated in Fig. 6.4(c).

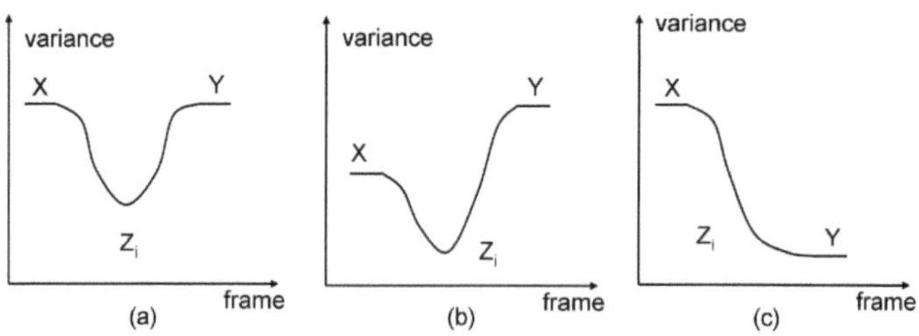

Fig. 6.4. The variance curves of some typical dissolve transitions.

The sub-FSM of the dissolve detector is designed to capture the characteristic curves shown in Fig. 6.4. A state variable, *AverageVariance*, is used for pinpointing the starting and ending frame of the dissolve transition. Its initial value is set to 3.5 and it is updated by the following IIR filter in state 0,

$$AverageVariance = AverageVariance \times 0.85 + HV_I \times 0.15 \tag{6.6}$$

where HV_I is the intensity histogram variance.

Verification is a key component of this FSM. The main challenge is that the variance curve may not be smooth due to motion or camera flashes in the original sequences *X* and/or *Y*. For verification purposes, a set of heuristic features are extracted based on the entire transition. In this section, we only introduce a few of them. For more details, please refer to [Liu06].

From the variance curve, shown in Fig. 6.5, the starting and ending frames need to be located. To do that, the system starts from the minimum variance frame in the candidate transition, and then searches forward and backward for the maximum absolute delta variance frames, which are f_{min} and f_{max} in the figure. Then from f_{min}, the system further searches backward until the delta variance of the current frame is less than half of the delta variance of the next frame or $2\times AverageVariance$. This frame is set as the starting frame of the candidate dissolve. Similarly, the system searches from f_{min} forward, and locates the ending frame.

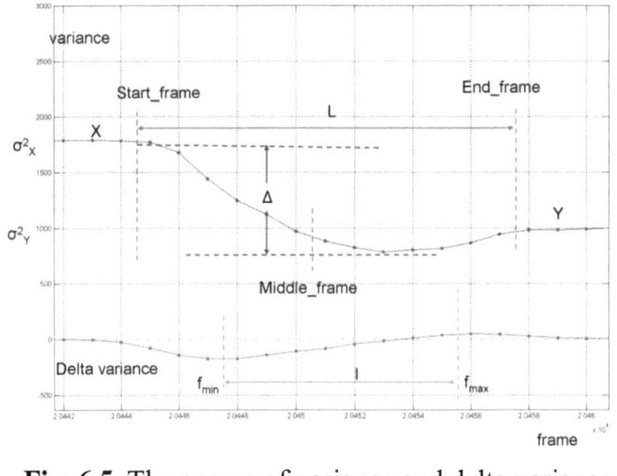

Fig. 6.5. The curves of variance and delta variance.

Then a set of heuristic features is extracted for verification purposes. For example, the height of the variance curve, Δ, is the difference of the maximum and minimum variances within the transition. Knowing that the delta variance is roughly a linear curve between f_{min} and f_{max}, a linear fitting is done for the delta variance. The system also computes the estimation error for each image in the transition from its neighboring images, and the matching error between the starting and ending frames of the transition.

The baseline dissolve verification employs a sequence of threshold-based criteria relying on these features. A more robust approach is to apply SVM on these features.

Wipe Detector

Wipe is the most ill defined transition. There are more than 20 different types of wipe that are commonly used in video editing and there is no single rule that applies to all of them. In this system, only one common type of wipe is considered, where the first scene gradually changes to the second scene, and for a given intermediate frame, part of the frame comes from the first scene, and part of it comes from the second scene.

A wipe is triggered by a smooth change, when the matching error ME_A is bigger than $1.5\times AverageME$ and less than $4\times AverageME$. In Fig. 6.6, the starting and the ending frames of the candidate wipe transition are denoted as X and Y, and an intermediate frame as Z_i, $i = 1, \ldots, L-1$, where L is the duration of the transition. The system partitions frame Z_i into 8×8 blocks, and finds the best match with motion compensation from both X and Y for each block. When the matching error is too high, the block does not come from either X or Y. Then the portion of blocks with a match from X, denoted as x_i, and the portion of blocks with a match from Y, denoted as y_i are computed. Finally, the system measures the linearity of x_i and y_i curves to verify the wipe transition.

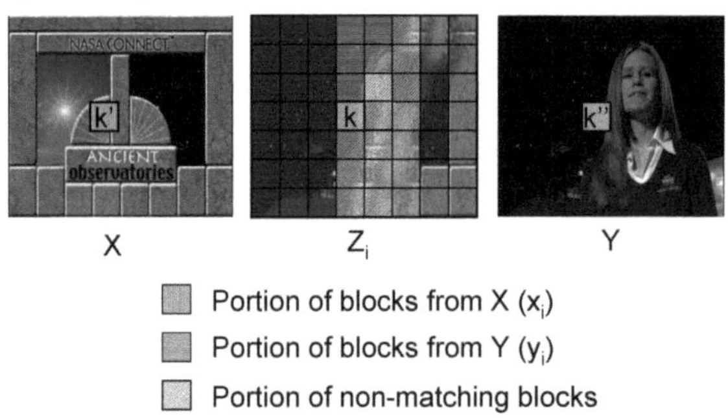

Fig. 6.6. Illustration of wipe verification.

SVM Models

Support vector machines are now a standard for fast and robust classification. While this classifier greatly reduces training time by analyzing only marginal samples, care must be given to the training parameters and underlying kernel selection. In [Liu06], the authors evaluated both linear and radial basis functions in a three-fold validation process. Seven linear settings and 70 radial basis function (RBF) settings were searched with random subsets of the training set split into 80/20 percent training/testing partitions. All features are globally normalized with a sigmoid before feeding the SVM.

6.2.3 Fusion of Detector Results

Fusion of detector results occurs when all frames are processed. The list of raw results is sorted by their starting frames and then all overlapped transitions are merged with different priorities assigned to each transition type. The adopted priority order is (from highest to lowest) FOI, dissolve, fast dissolve, cut, and wipe. The final step is to map the system types into two categories: cut and gradual. All shot boundaries except cuts are mapped into gradual.

6.2.4 Evaluation Results

In the TRECVID SBD evaluation, each group can submit up to 10 runs. Figure 6.7 shows the overall performance of all participants, with AT&T's runs plotted with squares. In terms of F-measure, the AT&T SBD system achieved the best overall performance. Table 6.1 shows the four best submissions for AT&T's SBD system in TRECVID2006.

Table 6.1 The best runs of AT&T's submissions.

Category	Performance (%)			Report localized changes	SVM Verification Kernel
	Recall	Precision	F-Measure		
Overall	**85.5**	**89.2**	**87.3**	No	Linear SVM
Cut	88.9	90.4	89.6		
Gradual	76.5	85.6	80.8		
Frame	87.1	91.9	89.4		
Overall	85.1	87.6	86.3	No	None
Cut	**89.4**	**90.4**	**89.9**		
Gradual	73.6	79.5	76.4		
Frame	86.9	93.0	89.8		
Overall	83.8	90.5	87.0	Yes	RBF 2
Cut	86.2	92.2	89.1		
Gradual	**77.5**	**85.8**	**81.4**		
Frame	87.4	92.3	89.8		
Overall	82.6	90.9	86.6	Yes	RBF 1
Cut	86.1	92.3	89.1		
Gradual	73.1	86.9	79.4		
Frame	**88.9**	**92.1**	**90.5**		

Among these results, there are different settings in terms of local changes and the inclusion of an SVM verification stage. The SVM based dissolve verification boosts the overall performance by 2.5% and gradual transition performance by 3.4%, a significant improvement when the initial performance is already high. The frame based gradual transition performance of all AT&T's 10 runs leads the other systems by more than 3.5%, meaning the proposed gradual transition (mainly the dissolves) boundary location approaches are very accurate. On an Intel 3.7GHz Xeon machine, all of the proposed system runs faster than 0.4× real time (the execution speed ranked the seventh among 26 participating groups).

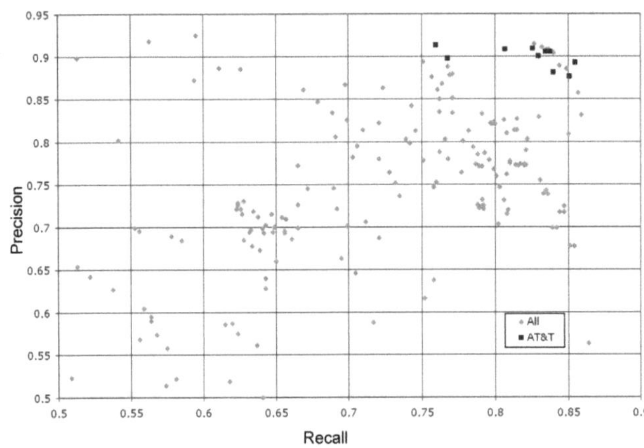

Fig. 6.7. SBD overall performance in TRECVID 2006.

6.3 Representative Image Selection

Each video shot can be represented by a set of representative images. There are a few ways to select the representative images. The simplest one is to choose the first, the last, or the middle frame of the shot to be the representative image. But this choice may not be optimal. Although the criteria of selecting optimal representative images are application dependent and subjective in most cases, there are some common standards. For example, in a video shot of weather forecast, images with front view faces and stable background are better than those with side view faces and fast motion background. Fig. 6.8 shows examples for these two cases. Obviously, the image shown in Fig. 6.8(a) is a better choice as a representative image for the shot. In this section, we briefly introduce a few methods of selecting representative image(s) within a shot.

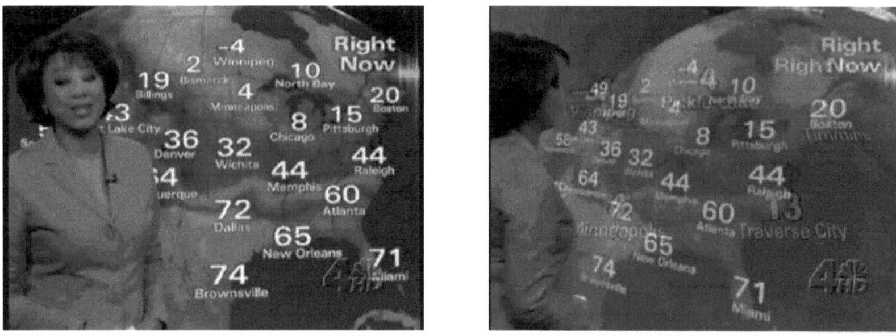

(a) Representative image (b) Non-representative image

Fig. 6.8. Two images within one weather forecast shot.

Two main factors that affect the determination of representative images are camera motions and local object motions. For stable shot with no camera motion and little object motion, the challenge of representative image selection is to choose one image with the best quality in terms of content. As mentioned before, when faces appear in the shot, it is preferred to choose images with frontal view faces, faces with opened eyes and closed mouth. If onscreen text appears in the shot, we want to choose one image with the most text and the top clarity.

Even when the camera is still, the scene may be complex and dynamic, moving foreground objects may obstruct the background or the major objects that the camera is focusing on, or the main objects may actually move in and out of view. Part of the shot may be out of focus because of the intrusion of new objects. The challenge in this scenario is to detect such obstructions by local motion detection or out of focus detection. Lim et al. proposed an algorithm that automatically determines if the captured photograph is out of focus through image analysis. It uses several global figures of merit which are computed from local image statistics [Lim05].

For videos with significant camera motions, including panning, tilting, zooming in, and zooming out, the scene within a shot may change completely. Multiple images are necessary to represent the shot. Detecting the types of camera motions and their exact occurrence frames is important to segment the shot into sub-shots where each sub-shot is significantly different from the previous sub-shot. Then, a representative image is assigned to each sub-shot. For example, an action of zooming in may separate the shot into two sub-shots, one containing the frames before the zooming in, and the other containing the frames after the zooming in action. For camera motion detection in compressed video, please refer to [Tan00].

In the following, we introduce a method to detect zooming in /out actions in a video proposed by Liu et al. [Liu07]. As shown in Fig. 6.9, frames i and $i-1$ are two adjacent frames. For each frame, the intensity values for the center row (horizontal bars h_i and h_{i-1}) and those for the center column (vertical bars v_i and v_{i-1}) are extracted. Dynamic programming is used to search the optimal match between the two horizontal bars, where the centers of the two bars are aligned. Fig. 6.9 shows an example of zooming out, and the best match path (MP_h) is marked in a solid line. The dotted solid line shows a possible match path for a case of zooming in. The tangent value of the angle of the match path (θ) is defined as the zooming factor. While zooming out, the factor is less than 1.0, and while zooming in, the factor is greater than 1.0.

Using the single pixel wide horizontal (vertical) bars, possible horizontal (vertical) zooming factors can be computed efficiently. Based on the optimal horizontal and vertical matching paths, the entire frames are used to verify the zooming decision. For the case of zooming out, frame i-1 is shrunk and compared to corresponding portion in frame i. The verification for the case of zooming in is similar. If the overall matching error is small enough, the zooming flag of current frame is set to be true, otherwise, false.

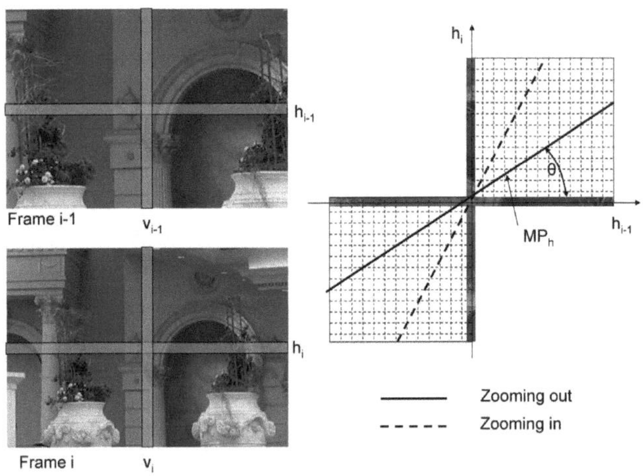

Fig. 6.9. Zooming detection.

Intelligently selecting the representative images benefits the video data management system, as a whole, and provides the end users with a more enjoyable and more informative video browsing and navigation experience.

6.4 Face Detection

Face detection in still images has a wide range of applications, including image retrieval, multimodal human computer interfaces, multimedia content analysis, digital cameras, face recognition, face tracking, video surveillance, emotion detection, etc. In the last two decades, much research effort has been focused on this area, and a few successful face detection techniques have been developed and adopted in real world applications and devices. For example, modern digital cameras feature face detection autofocus capabilities and recent Web cameras can enhance the video quality by adjusting the luminance around the faces. Recent progress in face detection algorithms can be found in two great review papers [Hjelmas01, Yang02]. In this section, we will give an overview of existing methods, and describe a couple of widely used approaches in detail.

Face detection is a challenging task, mainly due to the variant lighting conditions, occlusions, face orientation and poses, and existence of other facial features, for example, glasses, beards, etc. There is not a single approach that works well for all scenarios. Yang et al. surveyed about 150 different face detection methods, and categorized them into four general methods: (1) knowledge-based methods; (2) feature-based methods; (3) template matching-based methods; and (4) appearance-based methods.

Knowledge-based methods depend on some basic knowledge of human faces. For example, a frontal face is normally symmetric, with two eyes, one nose, and one mouth. If we project a frontal face vertically, the horizontal profile is normally symmetric, with a maximum value in the center, and decreasing values on both sides. If we project a frontal face horizontally, the vertical profile normally shows the low intensity valleys around the areas of eyes and mouths. Figure 6.10 shows the horizontal and vertical profiles of a typical face.

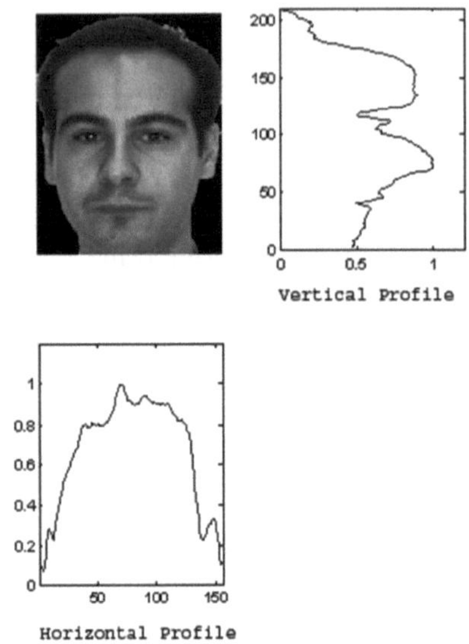

Fig. 6.10. Horizontal and vertical profiles of a typical face image.

Feature-based methods try to detect invariant features that are not significantly affected by lighting conditions, viewpoint, or pose. Examples of these features include facial features, texture, skin tone color, and multiple features. As an example, we show how to locate faces based on skin tone color. To effectively model skin color, we use the Hue Saturation Value (HSV) color system. Compared with standard Red Green Blue (RGB) color coordinates, HSV produces a more concentrated distribution for skin color. Most humans, despite the race and age, have similar skin hue, even though they may have different saturation and values. As value more depends on image acquisition setting, normally only hue and saturation are used to model human skin color. Fig. 6.11 gives the distribution of 2000 training data points, in hue-saturation space, that are extracted from different face samples. Clearly, it is appropriate to use a Gaussian with full covariance matrix to model this distribution. The hue of skin-color centroid is about 0.07, indicating that skin color is somewhat between red and yellow.

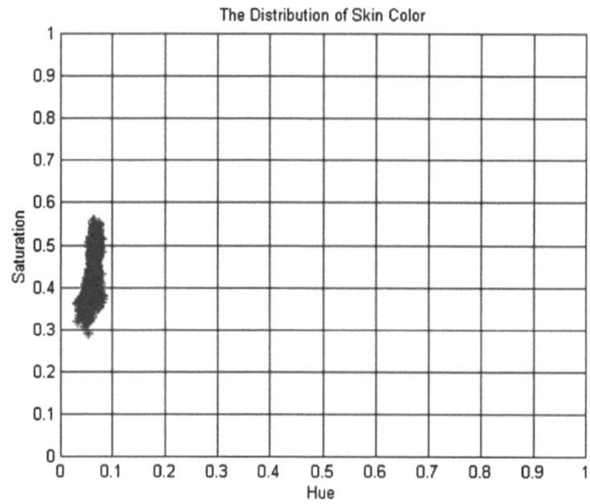

Fig. 6.11. Chroma chart of skin tone.

Figure 6.12 shows the skin tone color detection in a real image. The original image with complex background is shown in Fig. 6.12(a). Fig. 6.12(b) illustrates the skin tone color likelihood value (normalized to gray-scale for demonstration purposes). A preset threshold is used to classify pixels into face tone or non face tone categories. Then a morphological operation is applied to remove noise. The detected skin tone region is shown in Fig. 6.12(c). The white dot on the left corresponds to hand regions in the background, the white area on the right corresponds to face. A further step is required to verify detected skin region is face.

(a) Original image (b) Skin tone value (c) Skin tone region

Fig. 6.12. Face detection using skin tone.

Template matching methods compute the correlation of an input image with an average face template learned from a set of human faces (usually frontal faces). The correlation value can be computed at the global level, where the whole face is considered, or at the local level, where the matching of eyes, nose, and mouth are done separately. One challenge of template matching methods is that the face deformation needs to be tolerated. Liu and Wang proposed a fast template matching procedure using iterative dynamic programming (DP) for face detection and tracking in [Liu00]. In

Fig. 6.13, F is the face template image of size $M \times N$, T is the test image of size $I \times J$, and each small block represents a pixel. The task is to find a region in the test image T that is best matched with the template by some warping functions that map the columns/rows in the region to those of the template. Both global and local constraints for the warping functions are utilized to limit the search space. The global constraint is that the height and width of the face in test image are no less than those of the face template, and no bigger than twice of the face template. For a given top-left pixel position, s, of a candidate region, the regions for which we need to examine are all rectangles that end at any pixel within the shaded area. The largest candidate region is illustrated by a bold rectangle in the figure. The local constraint is that one or two rows/columns in candidate region are mapped to each row/column of the face template. An iterative 1-D dynamic programming procedure for row- and column-wise template matching is applied, and it converges to a local optimal 2-D matching.

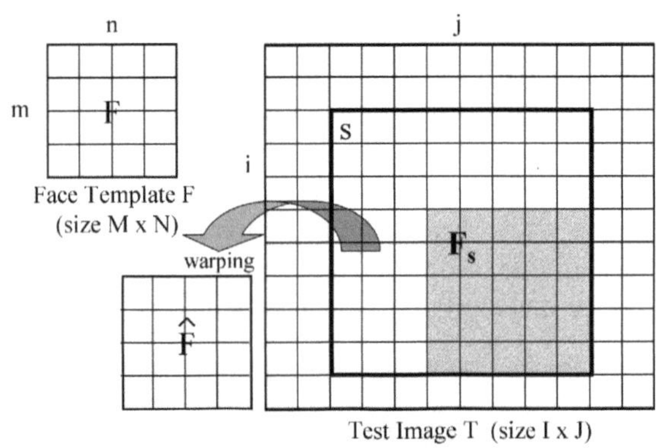

Fig. 6.13. Illustration of template matching.

Appearance-based methods rely on statistical analysis and machine learning techniques to capture the characteristics of face and non-face images. Except for selecting the specific models (neural network, eigenface, etc.), this is a purely data-driven approach. A large sized labeled dataset of face and non-face samples is required. In the following, we describe two successful face detection algorithms in this category.

Rowley et al. proposed a neural network based face detection algorithm in [Rowley98], where a 20×20 region is fed into a neural network after light correction and histogram equalization. The structure of the neural network is elaborately designed with three types of hidden neurons aiming to detect different facial features. This algorithm has been successfully employed in the Name it [Satoh99] project.

Viola and Jones provided an AdaBoost based face detection algorithm in [Viola01]. There are three main novelties in this algorithm. First, the authors introduced a new image representation called the "integral image," which allows very fast feature computation at many scales. The total number of these features is very large, much more than the number of pixels in the image. How to select a small set of effective features is actually the second novelty of this paper. Following the framework of AdaBoost, a weak learner is constructed based on only a single feature. While the boosting process selects a new weak classifier at each stage, effectively one new feature is selected. The third novelty is that the authors employed a method for combining successively more complex classifiers in a cascade structure which dramatically increases the speed of the face detector by focusing attention on promising regions of the image. The algorithm is very efficient, being able to detect faces in real time. The framework is flexible, and it can be used to detect other objects, for example, pedestrians. This algorithm has been implemented in the open source computer vision library (OpenCV) [OpenCV07].

It is more effective to combine these approaches to cope with the huge variation in visual appearance of faces. Schneiderman and Kanade [Schneider00] proposed a statistical method for face and car detection. They used a view-based approach with multiple detectors that are trained for specific orientations of an object. Each view-based detector models the faces using a product of histograms, each representing the joint statistics of a subset of wavelet coefficients and their position on the face. AdaBoost algorithm is used to minimize the classification error. Simulation results show that high accuracy is achieved for both frontal and profile face detections, although the speed is a bit slow – taking 1 minute to detect faces for a 320×240 image.

Face tracking is basically detecting faces in a sequence of images. Normally, the face detection in a still image involves searching faces in different locations with different scales, which is time consuming. In face tracking tasks, the face detection results of neighboring frames can be used for either speeding up the face detection in the current frame or improving the face detection accuracy. For example, if it is known that there is one face in the video and we detect a face in one frame, for the next frame, we only need to detect faces around the area that faces showed up before. The faces detected in one frame can be used to update the face model on the fly. Another advantage of face tracking in video is that motion information is a useful cue to locate possible face regions, which will significantly reduce the face detection complexity.

6.5 Face Recognition

The goal of face recognition is to recognize people by their facial features. Motivated by its wide area of applications, which include entertainment, human computer interfaces, law enforcement, biometric identification, and surveillance, etc., face recognition has attracted huge research attention and effort in the last few decades. Some of the advanced face recognition technologies, for example, eigenface based algorithms have been successfully applied in commercial products. Face recognition provides useful cues for video content analysis. In this section, we will discuss the approaches and challenges in the face recognition area. Interested readers can find more details on this topic in two excellent survey papers [Chellappa95, Zhao03].

Normally, a face detection system is composed of three main components: face detection, facial feature extraction, and face recognition. The last section focuses on face detection; this section will mainly describe the last two components. Face detection is a challenging problem because we must recognize the same person with different appearances, for example, different expressions and different hair styles.

There have been many face recognition algorithms proposed over the last 30 years. Zhao et al. classified them into three classes: (1) holistic matching methods, (2) feature-based (structural) matching methods, and (3) hybrid methods. Holistic matching methods treat the entire face region as input to the recognition system. Representative approaches within this category include the eigenface method and Support Vector Machine (SVM) based methods, etc.

Feature-based methods use local facial features such as eyes, nose and mouth and their locations and local statistics as input to the face recognition engine. This category includes pure geometric methods, dynamic link architectures, hidden Markov models, convolution neural networks, etc. We use convolution neural network as an example here.

Hybrid methods combine both local and global information about the face to recognize faces. Typical approaches are modular eigenfaces, hybrid local feature methods, component based methods, etc.

The method known as Eigenface was motivated by principal component analysis, which is a technique used to reduce multidimensional data to lower dimension for analysis or storage. It is also named the Karhunen-Loeve transform. This method requires a set of training images which are multiple images of each candidate that have been centered on the face and close cropped. Let the resolution of face image be $N \times N$, then the training set of faces is represented by a set of face image vectors $\{\mathbf{\Gamma}_1, \mathbf{\Gamma}_2, ..., \mathbf{\Gamma}_M\}$, where each face vector $\mathbf{\Gamma}_i$ is of length N^2. The average face vector is

$$\mathbf{\Psi} = \frac{1}{M}\sum_{i=1}^{M}\mathbf{\Gamma}_i \tag{6.7}$$

The difference between face $\mathbf{\Gamma}_i$ and $\boldsymbol{\Psi}$ is $\mathbf{\Phi}_i = \mathbf{\Gamma}_i - \mathbf{\Psi}$. Applying principal component analysis, we can find a set of M eigenvectors $\boldsymbol{u}_k$ and their associated eignevalues λ_k. They are the eigenvectors and eigenvalues of the covariance matrix:

$$\mathbf{C} = \frac{1}{M}\sum_{i=1}^{M}\mathbf{\Phi}_n\mathbf{\Phi}_n^T = \mathbf{A}\mathbf{A}^T, \quad where\ \mathbf{A} = [\mathbf{\Phi}_1\,\mathbf{\Phi}_2 \cdots \mathbf{\Phi}_M] \tag{6.8}$$

Matrix $\boldsymbol{C}$ is $N^2 \times N^2$, and determining the N^2 eigenvectors and eigenvalues is not manageable for even small image size. When the number of training face images, M is much smaller than N^2, there is a more efficient way to do so. Considering the eigenvectors $\boldsymbol{v}_i$ of A^TA, such that,

$$\mathbf{A}^T\mathbf{A}\mathbf{v}_i = \mu_i\mathbf{v}_i \tag{6.9}$$

where μ_i is the associated eigenvalue. It is obvious that $\boldsymbol{Av}_i$ are the eigenvectors of $\boldsymbol{C}=\boldsymbol{AA}^T$. Specifically, eigenvector $\boldsymbol{u}_i$ can be written as,

$$\mathbf{u}_i = \sum_{k=1}^{M}\mathbf{v}_{ik}\mathbf{\Phi}_k \tag{6.10}$$

The eigenvectors $\boldsymbol{u}_i$ are also called the eigenfaces. In practice, there is no need to use all M eigenfaces. [Turk91] found that $M'=40$ is sufficient to describe the set of face images. For a test face image $\boldsymbol{\Gamma}$, its eigenface components are computed by

$$\omega_k = \mathbf{u}_k^T(\mathbf{\Gamma} - \mathbf{\Psi}), \quad k = 1,...,M'. \tag{6.11}$$

The weight vector $\boldsymbol{\Omega}=\{\omega_1, \omega_2, ..., \omega_{M'}\}$ represents the contribution of each eigenface in describing the test face image. The simple method to determine the identification of the test face image is to find the face class k that minimizes the Euclidian distance

$$\varepsilon_k^2 = \left\|\mathbf{\Omega} - \mathbf{\Omega}_k\right\|^2 \tag{6.12}$$

where $\boldsymbol{\Omega}_k$ is the weight vector of the k-th face class.

Another approach for face recognition is by utilizing an artificial neural network. Typical images normally have thousands of variables (pixels), and a fully-connected feed-forward network needs hundreds of thousands of weights. When the training data is scare, overfitting problems may be occur. LeCun and Bengio proposed the so called convolutional networks for a wide range of applications including images, speech, and time series. Convolutional networks take advantage of three architectural ideas to ensure shift and distortion invariance as well as generalization ability. They are (1) local receptive fields, (2) shared weights, and (3) spatial or temporal subsampling. Fig. 6.14 shows a typical convoloutional neural network. Each unit of a layer connects to a set of units located in a small neighborhood in the previous layer. With local receptive fields, elementary visual features including edges can be extracted by neurons. To extract the same visual feature, neurons at different locations can share the same connection structure with the same weights. The output of such a set of neurons is a feature map. This operation is the same as a convolution of the input image with a small size kernel. Multiple feature maps can be applied to extract multiple visual features across the image. Subsampling is used to reduce the resolution of the feature map, and hence reduce the sensitivity of the output to shifts and distortions.

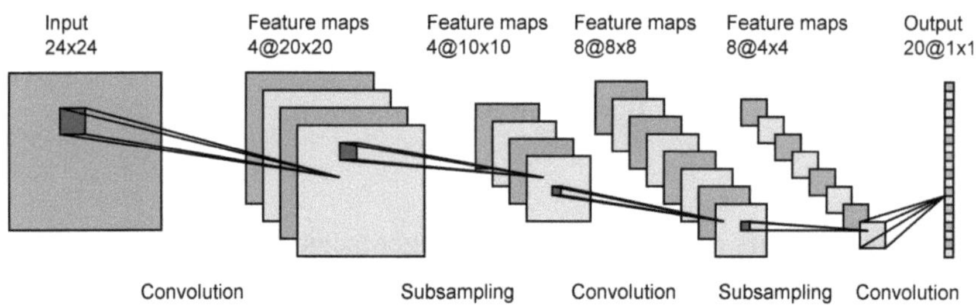

Fig. 6.14. A typical convolutional neural network for face recognition.

Lawrence et al. [Lawrence97] applied the convolutional neural network [LeCun98] approach to face recognition. The authors conducted experiments on the ORL database which was created by the Olivetti Research Laboratory in Cambridge, UK. There are 40 distinct subjects in this database, and each one has 10 different images. The experiment used 5 images for training and 5 for testing for each subject. The authors adopted a similar architecture of a convolutional neural network as shown in Fig. 6.14. The dimensions of the input images and all feature maps in the middle layers are different. Since there are 40 subjects to recognize, the output layer has 40 neurons. The best achieved error rate is 3.8% in this case.

Blanz [Blanz06] tackles the face detection problem using a 3D morphable model, which extracts complete shape and texture estimates as invariant facial features. Morphable model can be used as a preprocessing tool for generating frontal views from non-frontal images as the inputs to the frontal face recognition systems. Another advantage of this method is that 3D face reconstruction can help to build a large variety of different views of faces, which are used as a training set for learning 2D features that are invariant to pose and illumination.

Wiskott et al. [Wiskott99] presented a face recognition system where the database contains one image for each person. The system utilizes image graphs to collapse the variance introduced by position, size, expression, and pose. Faces are represented by labeled graphs, where edges are labeled with distance information and nodes are labeled with Gabor wavelet coefficients, called jets. The nodes correspond to the fiducial points (e.g. eyes, mouth, contour, etc.). After extracting the model graphs for both gallery images and probe images, face recognition is carried out by comparing the probe image graphs and all gallery image graphs.

6.6 Video Optical Character Recognition

Optical character recognition (OCR) is a technique to translate images of handwritten or printed text into symbols. As a field of research, OCR benefited from the progress in pattern recognition, artificial intelligence, and computer vision. Over the last half contrary, OCR techniques have been constantly improved, and many commercial products have been successfully built. The challenges of recognizing typewritten text are mainly image defects, similar symbols, punctuation, and various fonts. Nowadays, the accuracy of recognizing typewritten Latin text exceeds 99%, and it is considered as a solved problem. For handwritten input, it is still an active research area. PDAs produced by Palm adopted the Graffiti recognition system, and the Tablet PCs running the Microsoft operating system recognize pen input.

Video OCR locates textual information in a sequence of video frames. With Video OCR, program information, reporters' and interviewees' names, location, sport scores, pricing information, contact information, and other on-screen descriptions and annotations become valuable data for precise navigation of video.

Compared with OCR for still images, there are two major challenges in Video OCR. First, the resolution of video is low, and the text embedded in video is small too. The size of a character in video may be less than 10×10 pixels. With such low resolution, the regular OCR algorithms may not work reliably. Second, the video normally has complex background. Regular printed documents have a uniform (for example, white) background, and document segmentation can be easily conducted. With complex video backgrounds, both segmentation and recognition are more difficult. Given the large data volume nature of video (e.g. for NTSC video, the frame rate is 29.97 frames per second), it is time consuming to conduct OCR on each individual frame. A preferred technique is to segment video into shots, where the content within each shot remains stable and the onscreen text does not change. Then based on all frames within a shot, a simplified and effective OCR method is devised.

Lienhart [Lienhart03] provided an excellent survey on the core concepts underlying the texture-based approaches to automatic detection, segmentation, and recognition of visual text in complex image and videos. Hua et al [Hua02] addressed how to deal with the complex background in video OCR. They utilize multiple frames that contain the same text to get all clear words from these frames in two steps: (1) use multiple frame verification to reduce text detection false alarms; (2) detect and joint every clear text block from those frames to form a clearer manmade frame, which is sent to the OCR engine.

Sato et al. [Sato98] tackled the video OCR problem for digital news archives. They proposed two effective methods to address these two challenges. First, a sub-pixel interpolation filter is applied to enhance the original image resolution by four times in both the horizontal and vertical directions. Assume the original image is $I(x, y)$, and the high resolution image is $L(x, y)$. Then, $L(4x, 4y) = I(x, y)$ for pixels whose coordinates are multiples of 4. For other pixels, the value of $L(x, y)$ is interpolated by the following formula,

$$L(x,y) = \frac{\sum_{(x_0,y_0)\in N(x,y)} d(x-x_0, y-y_0)\cdot I(x_0/4, y_0/4)}{\sum_{(x_0,y_0)\in N(x,y)} d(x-x_0, y-y_0)} \tag{6.13}$$

where

$$N(x,y) = \{(x_0, y_0) \mid x_0 \in \{\lfloor x/4 \rfloor \cdot 4, \lceil x/4 \rceil \cdot 4\}, y_0 \in \{\lfloor y/4 \rfloor \cdot 4, \lceil y/4 \rceil \cdot 4\}\} \tag{6.14}$$

and $d(x,y) = \|(x,y)\|^{-1}$

To cope with the complex background issue in video, Sato et al. relied on the following observations. In video, the position of video captions is relatively stable across frames while there are usually motions in complex backgrounds. Captions usually have high intensity values such as white pixels, and a character normally consists of four different directional line elements: vertical, horizontal, left diagonal, and right diagonal. These observations inspired two techniques to deal with the challenge of complex backgrounds. First, a time-based minimum pixel value search was employed to enhance the text region and smooth out the variation of the background scene. Specifically, for a group of $n + 1$ frames $\{L_i(x, y), \ldots, L_{i+n}(x, y)\}$ that starts from frame i and ends at frame $n+i$, the filtered image $L_i'(x, y)$ is computed via $L_i'(x, y) = \min\{L_i(x, y), \ldots, L_{i+n}(x, y)\}$. Second, four directional line filters are applied to the image, and the output is integrated as a preprocessing of the input image.

A conventional correlation based pattern matching technique was used to recognize characters in video after the above mentioned processing. Evaluated on seven 30 minute CNN Headline News videos, the proposed method achieved a recognition rate of 83.5%. Although this is not comparable to the performance of regular OCR, it is very valuable for video indexing and searching.

Commercial products for video OCR are available on the market. Virage's on-screen text recognition plug-in provides real-time, automatic detection and identification of on-screen characters and numbers that appear in the video frame. Using ConTEXTract™ text extraction and recognition technology from SRI International®, this plug-in also allows users to focus VideoLogger on regions of interest in the video frame, for example the lower third for city or reporter names, or on an upper quadrant for time clock display.

6.7 Concept Detection

Traditional image search leverages text associated with images, a low level content-based matching, or a combination of the two. A more intuitive content search needs to facilitate semantic concepts. The signal processing community has long studied low level features and derived high-level features (or semantic concepts) for large image databases. High-level concepts are generally learned using patterns discovered over a set of images, where machine learning techniques are used to create discrete classifiers and provide deterministic scores for concept similarity.

The Disruptive Technology Office (DTO) sponsored the Large-Scale Concept Ontology for Multimedia (LSCOM) workshop to develop an expanded multimedia concept lexicon on the order of 1000 [LSCOM06].

Concepts related to events, objects, locations, people, and programs have been selected following a multi-step process involving input solicitation, expert critiquing, comparison with related ontologies, and performance evaluation.

Tsinghua's research team proposed to use the multi-label multi-feature learning (MLMF learning) [Yuan07] for concept detection. This approach warps the labeling information of many concepts with many features to learn a joint-concept distribution on the regional level as an intermediate representation. IBM research's approach for concept detection is to apply supervised learning algorithm to a set of low-level features [Campbell07]. These features include color histogram, color correlogram, color moments, co-occurrence texture, wavelet texture grid, edge histogram, and the locally normalized histogram of oriented gradient (HOG). The learning algorithms include support vector machines, subspace bagging, cross-domain learning and so on with different fusion strategies and cross-concept learning components for leveraging multi-modal and multi-concept relationship. Worring et al. developed MediaMill semantic search engine [Worring07], which computes a large lexicon of 491 concepts. The system defines a visual similarity space, a semantic similarity space, a semantic thread space and a few browsers, includes the Crossbrowser, the Spherebrowser, the RotorBrowser, and the GalaxyBrowser for the users to effectively explore the video collection.

Zavesky et al. [Zavesky07] reported their work on searching visual semantic spaces with concept filters. 374 concept classifier models (derived from the LSCOM lexicon [LSCOM06]) were trained over data in the TRECVID 2005 development set. TRECVID provides participants with a large collection of video data over which experiments are conducted and presented in an annual workshop. The training data is a set of over 64k TRECVID keyframes with overlapping positive and negative labels. The authors extracted low-level features for three modalities: grid-based color moments, a global edge direction histograms, and global Gabor texture responses. Classifier models for each of the 374 concepts were trained with support vector machines (SVMs) for the three feature modalities. The final score for each concept is computed as a weighted summation of the outputs of the three component classifiers. Thus, the final output of concept computation is a vector of 374 high-level concept scores for each keyframe in a video.

Here we briefly describe the visual features presented in [Zavesky07].

6.7.1 Color Feature

Color moment denotes the color distribution by using mean, standard deviation, and the third root of the skewness of each color channel [Stricker95]. To capture the special information, the image is divided into grids with size of $M \times N$. Within each grid, we can compute the color moments.

Let the i-th channel of an input image be $I^i(x, y)$, and the j-th grid covers a rectangle area specified by the left and right boundary in the x coordinate $\{x^j, x^j+M\}$ and the bottom and top boundary in the y coordinate $\{y^j, y^j+N\}$. Then color moments can be computed by

$$E_j^i = \frac{1}{MN} \sum_{x=x^j}^{x^j+M-1} \sum_{y=y^j}^{y^j+N-1} I^i(x,y), \tag{6.15}$$

$$\sigma_j^i = \left\{ \frac{1}{MN} \sum_{x=x^j}^{x^j+M-1} \sum_{y=y^j}^{y^j+N-1} \left[I^i(x,y) - E_j^i \right]^2 \right\}^{\frac{1}{2}},$$

$$s_j^i = \left\{ \frac{1}{MN} \sum_{x=x^j}^{x^j+M-1} \sum_{y=y^j}^{y^j+N-1} \left[I^i(x,y) - E_j^i \right]^3 \right\}^{\frac{1}{3}}$$

In [Zavesky07], CIE LUV color space is considered, and a 5×5 grid is used.

6.7.2 Texture Feature

Texture information is useful for visual concept detection. Manjunath and Ma [Manjunath96] used Gabor wavelet features for texture analysis and achieved good results. A 2-D complex Gabor function $g(x, y)$ is given by,

$$g(x,y) = s(x,y) w(x,y), \tag{6.16}$$

where

$$s(x,y) = \exp(2\pi j W x)$$

$$w(x,y) = \frac{1}{2\pi\sigma_x\sigma_y} \exp\left[-\frac{1}{2}\left(\frac{x^2}{\sigma_x^2} + \frac{y^2}{\sigma_y^2} \right) \right]$$

$s(x, y)$ is a complex sinusoidal, known as the carrier, and $w(x, y)$ is a 2-D Gaussian-shaped function known as the envelope. The Fourier transform of $g(x, y)$, $G(u, v)$ is

$$G(u,v) = \exp\left\{-\frac{1}{2}\left[\frac{(u-W)^2}{\sigma_u^2} + \frac{v^2}{\sigma_v^2}\right]\right\} \tag{6.17}$$

where σ_u=1/2$\pi\sigma_x$, and σ_v=1/2$\pi\sigma_y$. Gabor functions form a complete but non-orthogonal basis set. A class of self-similar functions, known as Gabor wavelets, can be derived from the mother Gabor wavelet $g(x, y)$,

$$\begin{aligned} & g_{mn}(x,y) = \\ & a^{-m} g\left(a^{-m}(x\cos\theta_n + y\sin\theta_n), a^{-m}(-x\sin\theta_n + y\cos\theta_n)\right), \\ & where \quad \theta_n = \frac{n\pi}{K}, \quad a > 1, \quad m,n \in Integer \end{aligned} \tag{6.18}$$

K is the total number of orientations. It is clear that $g_{mn}(x, y)$ is a scaled (by a^{-m}) and rotated (by θ_n) version of $g(x, y)$. The Gabor wavelets provide a localized frequency analysis.

For texture analysis, a subset of $g_{mn}(x, y)$, is used, which covers a range of frequencies (from U_l to U_h) at certain scales (S) with a certain number of orientations (K). A strategy for designing this subset is to ensure that the half-peak magnitude support of the filter responses in the frequency domain touch each other.

The Gabor wavelet transform for an image $I(x, y)$ is defined as

$$\begin{aligned} & W_{mn}(x,y) = \iint I(x',y') g_{mn}^{*}(x-x',y-y')dx'dy', \\ & m = 1,\ldots,S; n = 1,\ldots,K. \end{aligned} \tag{6.19}$$

where * denotes the complex conjugate. The mean and the standard deviation of the transform coefficients can be used to represent the texture information:

$$\mu_{mn} = \iint |W_{mn}(x,y)| dxdy, \quad \sigma_{mn} = \sqrt{\iint \left(|W_{mn}(x,y)| - \mu_{mn}\right)^2 dxdy} \tag{6.20}$$

[Zavesky07] used a combination of four scales and six orientations to extract texture features.

6.7.3 Edge Feature

An edge direction histogram denotes the distribution of edge directions. The number of bins is 73, which includes 72 bins for edge direction quantized at 5 degrees and one bin is for non-edge points. The Canny filter is applied to detect edge points, and for each detected edge point, we calculated its gradient by a Sobel operator. This histogram is normalized by the number of all pixels to cancel the effect of the image size.

A support vector machine (SVM) is employed to map the low level features to concepts. The state of the art of concept detection is far from perfect. TRECVID 2007 reported the best run on concept detection achieved an inferred average precision (infAP) of 0.13. Even though the concept detection accuracy is not reliable, using concept filters to search and browse video content is still very promising [Zavesky07] and for particular concepts the performance is much better.

6.8 Video Browsing

An efficient video browsing interface is important for a complete video content management system. Generally, we browse videos at three different levels. At the first level, we glance at a set of video clips, either as search results or a group of videos with similar theme. At this level, we normally look at the thumbnails extracted from video or brief descriptions of the clip and then choose one or two to dive in. At the second level, we skim the content within a single video. At this level, there is still intensive user interaction to select the right story to view more details. Depending on the interface, viewing at this level is normally nonlinear, and the user interface may provide an efficient navigation capability to ease the task of locating a specific story. At the third level, the user finds the right video segment, and consumes the video linearly. Not much interaction is involved, but at this level, side information can be composed together with the original video content to enhance the browsing experience. Certain fast playback functions, like fast forward, jumping to the next shot, etc., give the user more control over the pace while watching video. In this section, we present a few different ways to present video content at these three levels.

Many different representations for browsing video have been proposed including "light-table" or "video mosaic" views which are two-dimensional arrays of thumbnail images, "story boards" which may include dialog text with thumbnails, super-resolution or composite still images as well as summarized videos. Further, where thumbnails are presented, animation or video summarization may also be used. Video mosaics present all the content related to a dedicated theme, such as sports, music, kids, query of certain topic, etc., on one page. The viewer can then navigate from one thumbnail to another, with synchronized audio or to an interactive service available on the themed portal (real time info, scores, games, etc.). Fig. 6.15 shows an example where multiple video clips are played back simultaneously, while the audio of the selected video is played back [httv07].

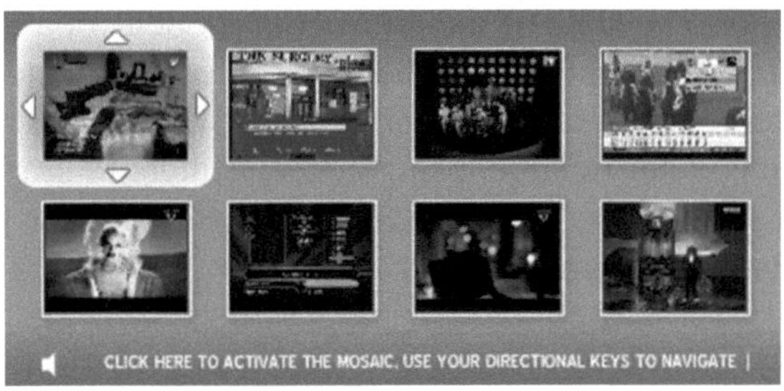

Fig. 6.15. Video Mosaic.

For recurring content, like the NBC Nightly News program, calendar view is a perfect choice to easily browse the video archive. Fig. 6.16 shows a calendar based interface developed in the MIRACLE system. Clicking any date will bring the user to the selected program directly.

Fig. 6.16. Calendar view of a video archive.

Chen et al. [Chen07] proposed a platform for showing the retrieved content in a geospatial and temporal manner. Fig. 6.17 illustrates the capability of the platform to search a corpus of collected video over a period of time and present an interesting aggregated RSS feed to the user. The figure shows the query for "Klose," the top scoring World Cup 2006 soccer player. In this scenario, a user is interested in following a player, namely Miroslav Klose, and viewing his goals throughout the tournament. The annotations in Fig. 6.17 are as follows:

1. Feed title – This appears in the media RSS document that links to the search engine.
2. Video clip – This is a link to a video stream corresponding to the search term (in this case, the first occurrence of the goal scoring video clip).
3. Thumbnail – This is a link to an image snapshot of one of the goals of Klose.
4. Audio clip – This is a link to the audio stream corresponding to the search term.
5. MediaRSS URL – This is a link to the media RSS (Really Simple Syndication) URL that is shown on the map.

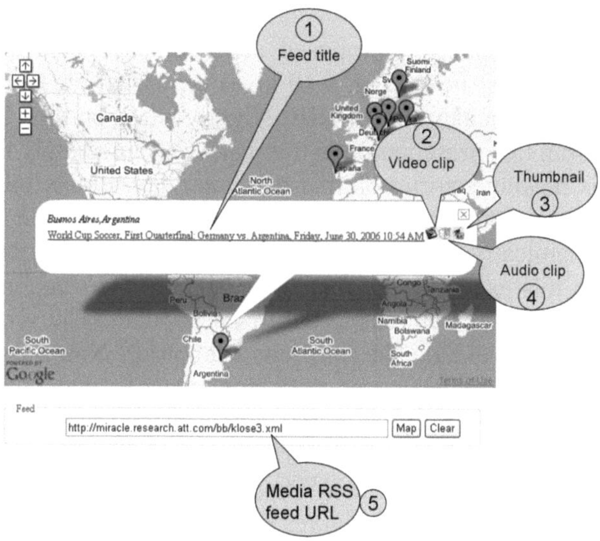

Fig. 6.17. Visualization from a corpus of 2006 FIFA World Cup data.

Fig. 6.18 is a view of a single video program which conveys the program content and permits rapid media browsing. The images are linked to the media as is each sentence. The "Program segment" control is used to sequentially navigate the document, and the extent of time corresponding to the current page is indicated as a gray bar on a timeline. The timeline spans the entire program length, and shows the locations of the query matches, with mouse-over indicating context. Clicking on an occurrence moves the view to the page containing the hit, and the query matches on the current page are highlighted in a different color.

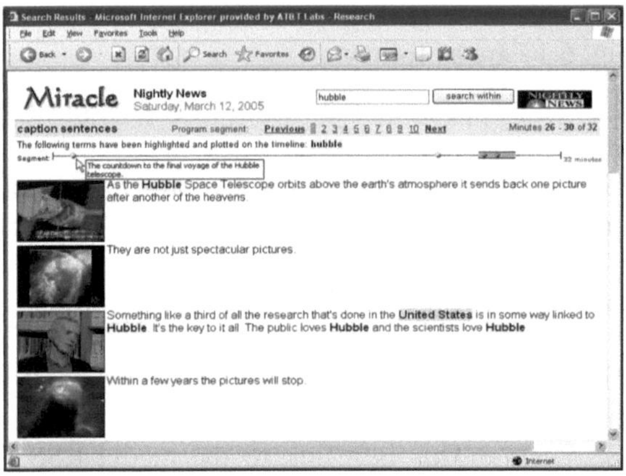

Fig. 6.18. Multimedia document page showing term and entity highlighting.

Huang et al. [Huang99] proposed an automatic semantic extraction method for broadcast news. Fig. 6.19 gives the visual presentation for the news summary of the day for the NBC Nightly News on the 12th of February, 1998. From this presentation, a user can see immediately that there are a total of six headline stories on that particular day. Below the representative image for each story, the list of its keywords is displayed as a right-to-left flow dynamically so that users can get a sense of the story from the keywords. In this example, the first story is about the weapons inspection in Iraq where Russians are suspected to tip Saddam. The second story is about the Clinton scandal. The third one is about El Nino. The fourth one is about whether secret service workers should testify against the president. The fifth is about the high suicide rate among youngsters in an Indian village. The sixth is about government's using tax dollars to pay the rent for empty buildings. Although some information is lost due to converting from a time varying display to this figure, from these examples, the effectiveness of this story-telling visual representation for the news summary is evident.

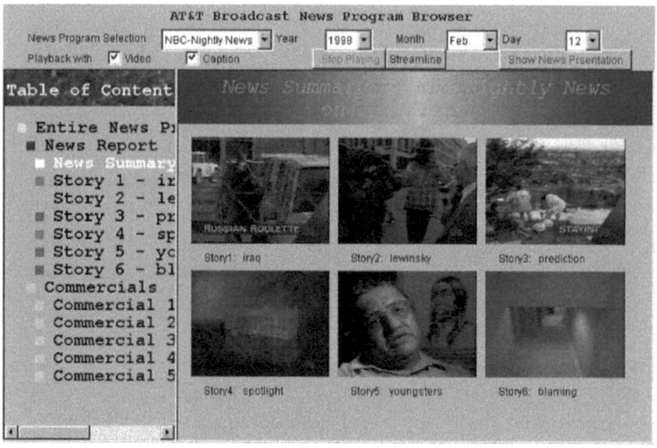

Fig. 6.19. Representation of news summary of a program.

Yeo and Yeung [Yeo97] proposed to use storyboard view to show the video content. Fig. 6.20 shows an example of such a story board. The visual layout and relative sizes of the sub-images reflect the relative dominance, or importance, of the segments they represent.

Fig. 6.20. Storyboard for a story.

Huang et al. [Huang99] presented a story based news presentation interface, shown in Fig. 6.21. The presentation for each story has three parts: the upper left corner is a set of 10 keywords automatically chosen from the segmented story based on the relative importance of the words; the right part displays the text of the story; the rest is the visual presentation of the story consisting of five images chosen from video in the content based manner described above. The example shown is the visual representation about the El Nino story.

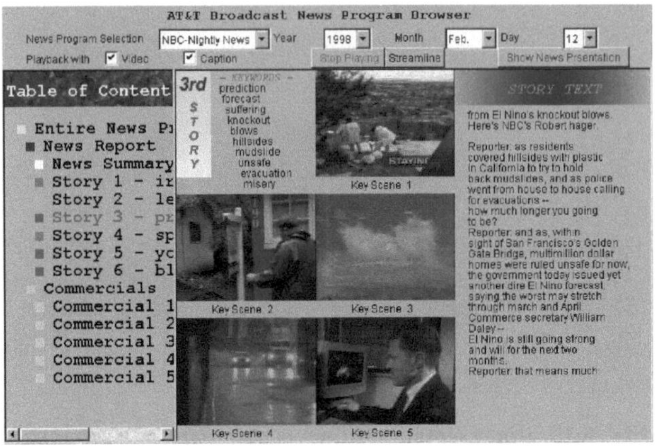

Fig. 6.21. Visual representation of a news story.

It is obvious that the story representation constructed this way is compact, semantically revealing, and visually informative with respect to the content of the story. A user can choose either to scroll the text on the right to read the story or to click on a thumbnail image to playback video from the corresponding keyframe. Compared with linear browsing or low level scene cut browsing, this system allows a more effective content based non-linear information retrieval.

6.9 Conclusion

With the booming of the World Wide Web and consumer electronics, enormous amounts of video are available for the users. Video data management becomes important and it requires more advanced video content analysis tools to provide more accurate video search and more effective video browsing. In this chapter, we introduced and discussed a few components of video content analysis. They are shot boundary detection, representative image selection, face detection and recognition, text extraction in video, concept detection, and video browsing and navigation. Semantics embedded in the video stream can be partially extracted from video using the presented technologies. To extract more information from video, we need more advanced and more reliable video processing tools and multimodal processing techniques.

References

[Blanz06] Blanz, V.: Face recognition based on a 3D morphable model. *Automatic Face and Gesture Recognition*, pp. 617–624 (2006).

[Campbell07] Campbell, M., Haubold, A., Liu, M., Natsev, A., Smith, J., Tesic, J., Xie, L., Yan, R., Yang, J.: IBM research TRECVID-2007 video retrieval system. *TRECVID 2007 Workshop* (2007).

[Chang07] Chang S.F., Kennedy, S.L., Zavesky, E.: Columbia University's semantic video search engine. *ACM CIVR* (2007).

[Chellappa95] Chellappa, R., Wilson, C.L., Sirohey, S.: Human and machine recognition of faces: A survey. *Proceedings of the IEEE,* **83**(5) pp. 705–741 (1995).

[Chen07] Chen, Y.F., Fabbrizio, G.D., Gibbon, D., Jana, R., Jora, S., Renger, B., Wei, B.: GeoTracker: geospatial and temporal RSS navigation. *WWW* (2007).

[Dimitrova02] Dimitrova, N., Zhang, H., Shahraray, B., Sezan, I., Huang, T., Zakhor, A.: Applications of video-content analysis and retrieval. *IEEE Multimedia*, **9**(3), pp. 42–55 (2002).

[Flickner95] Flickner, M., Sawhney, H., Niblack, W., Ashley, J., Huang, Q., Dom, B., Gorkani, M., Hafner, J., Lee, D., Petkovic, D., Steele, D., Yanker, P.: Query by image and video content: The QIC System. *IEEE Computer,* **38**, pp. 23–31 (1995).

[Flickr07] Flickr, http://www.flickr.com, cited 10 Dec. 2007.

[Gibbon06] Gibbon, D., Liu, Z., Shahraray, B.: The MIRACLE video search engine. *IEEE CCNC* (2006).

[Gonzalez93] Gonzalez, R.C. and Woods, R.E.: *Digital Image Processing*. Addison Wesley (1993).

[Hjelmas01] Hjelmas, E. and Low, B.K.: Face Detection: a Survey. *Computer vision image understanding*, **83**(3), pp. 236–274 (2001).

[httv07] Video Mosaic System, High Tech TV, http://www.httv.fr /pages/prodsandsols/videomosaic.html, cited 10 Dec 2007.

[Hua02] Hua, X., Yin, P., Zhang. H.: Efficient video text recognition using multiple frame integration. *ICIP* (2002).

[Huang99] Huang, Q., Liu, Z., Rosenberg, A., Gibbon, D., Shahraray, B.: Automated semantic structure reconstruction and representation generation for broadcast news. *SPIE* (1999).

[Kraaij06] Kraaij, W., Over, P., Smeaton, A.: TRECVID 2006 – An introduction. *TRECVID 2006 Workshop* (2006).

[Lawrence97] Lawrence, S., Giles, C., Tsoi, A., Back, A.: Face recognition: a convolutional neural network approach. *IEEE Transactions On Neural Networks*, **8**(1), pp. 98–113 (1997).

[LeCun98] LeCun, Y. and Bengio, Y.: Convolutional networks for images, speech, and time series. The Handbook of Brain Theory and Neural Networks, MIT Press, pp. 255–258 (1998).

[Lienhart03] Lienhart, R.: Video OCR: A survey and practitioner's gudie. *Video Mining*, Kluwer Academic Publisher, pp. 155–184 (2003).

[Lim05] Lim, S.H., Yen, J., Wu, P.: Detection of out-of-focus digital photographs. HP Labs Report HPL-2005-14, http://www.hpl.hp.co.uk/techreports/2005/HPL-2005-14.pdf, cited 10 Dec 2007.

[Liu00] Liu, Z. and Wang, Y.: Face detection and tracking in video using dynamic programming. *ICIP* (2000).

[Liu06] Liu, Z., Gibbon, D., Zavesky, E., Shahraray, B., Haffner, P.: AT&T research at TRECVID 2006. *TRECVID 2006 Workshop* (2006).

[Liu07] Liu, Z., Zavesky, E., Gibbon, D., Shahraray, B., Haffner, P.: AT&T research at TRECVID 2007. *TRECVID 2007 Workshop* (2006).

[LSCOM06] LSCOM Lexicon Definitions and Annotations Version 1.0. DTO Challenge Workshop on Large Scale Concept Ontology for Multimedia. Columbia University ADVENT Technical Report #217-2006-3 (2006).

[Manjunath96] Manjunath, B.S. and Ma, W.Y.: Texture features for browsing and retrieval of image data. *IEEE Transactions on Pattern Analysis and Machine Intelligence*, **18**(8), pp. 837–842 (1996).

[Martinez02a] Martinez, J.M., Koenen, R., Pereira, F.: MPEG-7: the generic multimedia content description standard, Part 1. *IEEE Multimedia*, **9**(2), pp. 78–87 (2002).

[Martinez02b] Martinez, J.M.: Standards – MPEG-7 overview of MPEG-7 description tools, Part 2. IEEE Multimedia. **9**(3), pp. 83–99 (2002).

[OpenCV07] Open Computer Vision Library, http://sourceforge.net/projects/opencvlibrary/, cited 10 Dec 2007.

[Ponceleon98] Ponceleon, D., Srinivasan, S., Amir, A., Petkovic, D., Diklic, D.: Key to Effective Video retrieval: Effective cataloging and browsing. *ACM Multimedia*, pp. 99–107 (1998).

[Rowley98] Rowley, H.A., Baluja, S., Kanade, T.: Neural network based face detection. *IEEE Transactions on Pattern Analysis and Machine Intelligence*, **20**(1), pp. 22–38 (1998).

[Sato98] Sato, T., Kanade, T., Hughes, E., Smith, M.A.: Video OCR for digital news archives. IEEE Workshop on Content-Based Access of Image and Video Databases. pp. 52–60 (1998).

[Satoh99] Satoh, S., Nakamura, Y., Kanade, T.: Name-it: Naming and detecting faces in news videos. *IEEE Multimedia Magazine*, **6**(1), pp. 22–35 (1999).

[Schneider00] Schneiderman, H. and Kanade, T.: A statistical method for 3D object detection applied to faces and cars. *CVPR* (2000).

[Shahraray95] Shahraray, B.: Scene change detection and content-based sampling of video sequences. Digital Video Compression: Algorithms and Technologies, *Proc. SPIE 2419* (1995).

[Stricker95] Stricker, M.A. and Orengo, M.: Similarity of color images. Proceedings of SPIE, Storage and Retrieval for Image and Video Databases III. **2420**, pp. 381–392 (1995).

[Tan00] Tan, Y-P, Saur, D.D., Kulkami, S.R., Ramadge, P.J.: Rapid estimation of camera motion from compressed video with application to video annotation. *IEEE Transactions On CSVT*, **10**(1), pp. 133–146 (2000).

[Turk91] Turk, M.A. and Pentland, A.P.: Face recognition using Eigenfaces. *Proc. CVPR*, pp. 586–591 (1991).

[Vapnik98] Vapnik, V.N.: *Statistical Learning Theory*. John Wiley (1998).

[Viola01] Viola, P. and Jones, M.: Robust real-time object detection. *IJCV* (2001).

[Virage07] Virage Products Overview, http://www.virage.com/content/products/index.en.html, cited on 10 Dec 2007.

[Wang00] Wang, Y., Liu, Z., Huang, J.: Multimedia content analysis using audio and visual information. *IEEE Signal Processing Magazine*, **17**(6), pp. 12–36 (2000).

[Wiskott99] Wiskott, L., Fellous, J., Kruger, N., and Malsburg, C.: Face recognition by elastic bunch graph matching. *Intelligent Biometric Techniques in Fingerprint and Face Recognition*, CRC Press, pp. 355–396 (1999).

[Worring07] Worring, M., Snoek, C., Rooij, O., Nguyen, G., and Smeulders, A.: The MediaMill semantic video search engine. *ICASSP* (2007).

[Xiao07] Xiao, Y., Du, X., Zhang, J., Hu, F., Guizani, S.: Internet protocol television (IPTV): The killer application for the next-generation Internet. *IEEE Communications Magazine*, **45**(11), pp. 16–134 (2007).

[Yang02] Yang, M., Kriegman, D., Ahuja, N.: Detecting faces in images: A survey. *IEEE Transactions On Pattern Analysis and Machine Intelligence*, **24**(1), (2002).

[Yeo95] Yeo, B.L. and Liu, B.: Rapid scene analysis on compressed video. *IEEE Transactions on Circuits and Systems for Video Technologies*, **5**(6), pp. 533-544 (1995).

[Yeo97] Yeo, B.L. and Yeung, M.M.: Retrieving and visualizing video. *Communications of the ACM*, **40**(12) (1997).

[YouTube07] YouTube, http://www.youtube.com/, cited 10 Dec 2007.

[Yuan07] Yuan, J., Z. Guo, et al: THU and ICRC at TRECVID 2007. *TRECVID 2007 Workshop* (2007).

[Zavesky07] Zavesky, E., Liu, Z., Gibbon, D., Shahraray, B.: Searching visual semantic spaces with concept filters. *ICIC* (2007).

[Zhang93] Zhang, H.J., Kankanhalli, A., Smoliar, S.W.: Automatic partitioning of full-motion video. *ACM Multimedia System*, **1**(1), pp. 10–28 (1993).

[Zhao03] Zhao, W., Chellappa, R., Rosenfeld, A., Phillips, P.J.: Face recognition: A literature survey. *ACM Computing Surveys*, pp. 399–458 (2003).

7 Audio Processing

7.1 Introduction

Audio plays an important role in our daily life. From speech to music, from FM radios to Podcast services, from lectures to audio books, audio is simply ubiquitous. Through audio, we sense the environment, acquire knowledge, exchange information, enjoy melodies, and so on. Nowadays, with the ease of audio creation, audio archiving, and audio distribution, the amount of audio data is unprecedented, and it far exceeds the capacity of individuals to consume. The value of audio data relies not only on its intrinsic merit, but also on how easy it is to access. Very often, we want to search for a piece of audio that we either heard before, for example, a specific song or a conference recording, or one that we are not aware of, for example, a speech of President Kennedy or a piece of country music. Obviously, metadata, including the title of the audio clip, the name of the producer, the category, a short summarization, etc. is extremely useful for searching audio. But in many cases, we desire an automatic mechanism to discover the audio content since the associated metadata is not sufficient. The same method can also enhance metadata based audio query approaches, because it is able to pinpoint the segment of interest within an audio clip more precisely. Content based audio indexing is a promising approach. With the support of content-based audio indexing and retrieval services, locating the desired audio information among a nearly infinite amount of audio data, is no longer a daunting task.

Audio content segmentation and categorization usually serve as the beginning steps in audio content analysis. Then, specific analysis methods can be utilized to process different types of audio, for example, speech and speaker recognition algorithms can be applied on speech signals, and note/pitch detection algorithms can be applied on music signals [Pfeiffer96]. Zhang and Kuo [Zhang00] addressed audiovisual data segmentation, indexing and retrieval based on multimodal content analysis, and content-based management of audio data. Baluja and Covell [Baluja07] presented an audio identification system, *Waveprint*, which creates compact fingerprints of audio data by combining computer vision techniques and large scale data stream processing algorithms. Thong et al [Thong03] presented a distributed multimedia content analysis and indexing architecture. The proposed architecture runs on low cost commodity hardware, and it is able to process large volumes of audio recordings with minimal support and maintenance. Its success is demonstrated by an audio and video search engine, *SpeechBot*. Lu and Hanjalic [Lu08] presented an unsupervised approach to discover key audio elements in general audio documents. The audio elements are treated as the text words in classical text document retrieval, and content based audio analysis and retrieval is enabled by applying the text analysis theories and methods.

With the mature of many multimedia content index and analysis technologies and the rapid growth of multimedia content, MPEG-7 [Chang01], formally named "Multimedia Content Description Interface," is a timely standard that allows search engines from different vendors to effectively identify multimedia content in large scale data sources. Kim et al discussed the details of MPEG-7 audio in their book [Kim06], where the interested readers can find more information about audio content description.

This chapter focuses on the fundamentals of audio analysis, the content-based audio processing and indexing technologies, and audio query and browsing methods. First, we introduce audio signal representation and some typical audio features in Sect. 7.2. Then, a set of commonly used audio features are described in Sect. 7.3. In Sect. 7.4, we show speaker segmentation and audio scene segmentation methods. Audio content categorization, including speaker recognition, audio scene detection, and music genre detection are addressed in Sect. 7.5. Then, the fundamentals of automatic speech recognition (ASR) are presented in Sect. 7.6. In Sect. 7.7, we illustrate a few audio query and browsing techniques. In this section, we discuss one-best word search, lattice and phonetic search, and query by example. Finally, our conclusion is drawn in Sect. 7.8.

7.2 Audio Signal and Its Representation

Figure 7.1 shows the waveform of a typical segment in a news broadcast. The audio signal carries rich information at different semantic levels. For a human being, we can easily tell speech from the other types of audio, for example, music, songs, or noise. Without understanding the linguistic meanings of the audio signal, we can also separate male speakers and female speakers. With a bit more effort, we may recognize the identities for known speakers, or judge weather two segments of speech are from the same speaker or not. As one of the most important communication mechanisms, speech signals obviously contain a wealth of linguistic information, which is very important for understanding the underlying content.

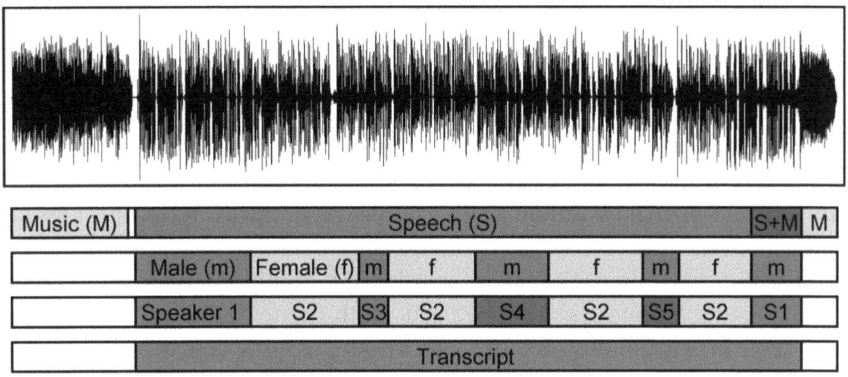

Fig. 7.1. Content of audio signal.

Audio signals can be represented in both the time and frequency domains. Generally speaking, audio signals are non-stationary, but within a short period, e.g. 5 - 50 milliseconds, the signal is relatively stationary. Within this short piece of audio, which is normally called a frame, we can analyze its spectrum. The spectrum is defined as the amplitude of the Fourier transform of the audio frame. Figure 7.2 shows the waveform and the spectrum of an audio frame which is 32 ms long. The sampling rate is 16 kHz, hence the audio frame contains 512 samples.

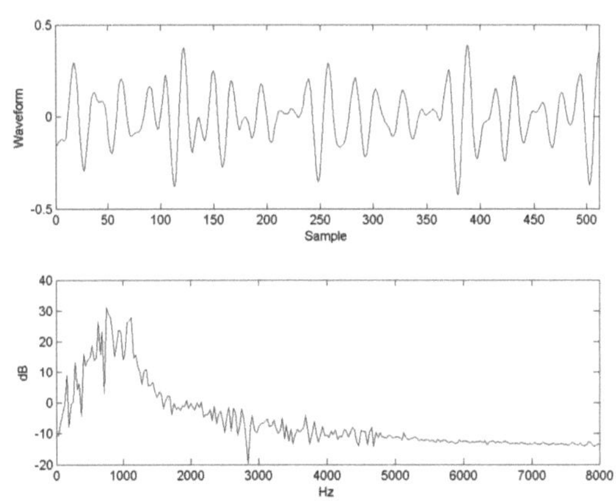

Fig. 7.2. Waveform and spectrum of an audio frame.

7.3 Audio Features

There are many features that can be used to characterize audio signals [Wold96, McKin03]. Usually audio features are extracted at two levels: short-term frame-level and long-term clip-level. The concept of the audio frame comes from traditional speech signal processing, where analysis over a very short time interval has been found to be most appropriate. For a feature to reveal the semantic meaning of an audio signal, analysis over a much longer period is necessary, usually from one second to several tens of seconds. Here we call such an interval an audio clip.

A clip consists of a sequence of frames and clip-level features that usually characterize how frame-level features change over a clip. The clip boundaries may be the result of audio segmentation such that the frame-level features within each clip are similar. Alternatively, fixed length clips, usually 1 to 2 seconds (s) may be used. Both frames and clips may overlap with their previous ones, and the overlapping lengths depend on the underlying application. Figure 7.3 illustrates the relationship of frames and clips. In the following, we first describe frame-level features, and then move onto clip-level features.

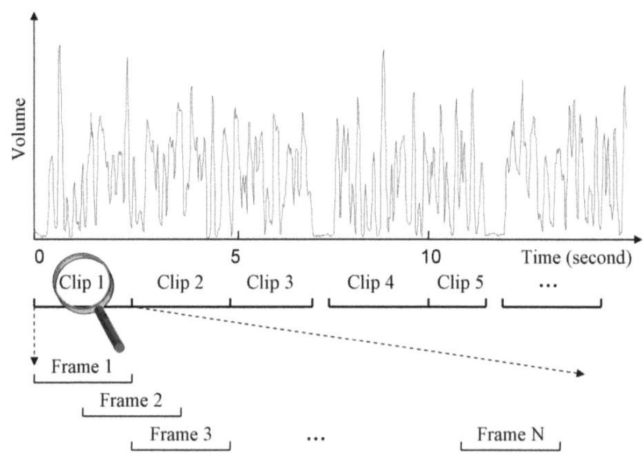

Fig. 7.3. Audio clips and audio frames.

7.3.1 Frame-Level Features

Most of the frame-level features are inherited from traditional speech signal processing. Generally they can be separated into two categories: time-domain features, which are computed from the audio waveforms directly, and frequency-domain features, which are derived from the Fourier transform of samples over a frame. Some features, for example, pitch and linear predicative code (LPC) coefficients, have an interpretation in both time and frequency domains. In the following, we use N to denote the frame length, and $s_n(i)$ to denote the i-th sample in the n-th audio frame.

Volume

The most widely used and easy-to-compute frame feature is volume. Volume is a reliable indicator for silence detection, which may help to segment an audio sequence and to determine clip boundaries. Normally volume is approximated by the root mean square (RMS) of the signal magnitude within each frame. Specifically, the volume of frame n is calculated by

$$v(n) = \sqrt{\frac{1}{N}\sum_{i=0}^{N-1} s_n{}^2(i)} \tag{7.1}$$

Note that the volume of an audio signal depends on the gain value of the recording and digitizing devices. To eliminate the influence of such device-dependent conditions, we may normalize the volume for a frame by the maximum volume of some number of previous frames.

Zero Crossing Rate

Besides the volume, zero crossing rate (ZCR) is another widely used temporal feature. To compute the ZCR of a frame, we count the number of times that the audio waveform crosses the zero axis. Formally,

$$Z(n) = \frac{1}{2}\left(\sum_{i=1}^{N-1}\left|sign[s_n(i)] - sign[s_n(i-1)]\right|\right)\frac{f_s}{N} \tag{7.2}$$

where f_s represents the sampling rate. ZCR is one of the most indicative and robust measures to discern unvoiced speech (i.e. the consonant sounds produced in the mouth not using the vocal chords). Typically, unvoiced speech has a low volume but a high ZCR. By using ZCR and volume together, one can prevent low energy unvoiced speech frames from being classified as silent.

Pitch

Pitch is the fundamental frequency of an audio waveform, and is an important parameter in the analysis and synthesis of speech and music. Normally only voiced speech and harmonic music have well-defined pitch. But we can still use pitch as a low-level feature to characterize the fundamental frequency of any audio waveform. The typical pitch frequency for human speech is between 50 and 450 Hz, whereas the pitch range for music is much wider. It is not easy to robustly and reliably estimate the pitch value for an audio signal. Depending on the required accuracy and complexity constraints, different methods for pitch estimation can be applied [Hess83].

One can extract pitch information by using either temporal or frequency analysis. Temporal estimation methods rely on computation of the short term autocorrelation function $R_n(k)$ or the average magnitude difference function (AMDF) $A_n(k)$, where

$$R_n(k) = \sum_{i=0}^{N-k-1} s_n(i)s_n(i+k) \quad and \quad A_n(k) = \sum_{i=0}^{N-k-1} \left|s_n(i+k) - s_n(i)\right| \tag{7.3}$$

Figure 7.4 shows the autocorrelation function and AMDF for a typical voiced speech sample. We can see that there exist periodic peaks in the auto-correlation function. Similarly, there are periodic valleys in the AMDF. Here peaks and valleys are defined as local extremes that satisfy additional constraints in terms of their values relative to the global minimum and their curvatures. For example, the AMDF in Fig. 7.4 has three valleys (marked with circles), and the pitch frequency is the reciprocal of the time period between the origin and the first valley. Similarly, there are three peaks in the autocorrelation graph in Fig. 7.4. Such valleys (and peaks) exist in voiced and music frames and vanish in noise or unvoiced frames.

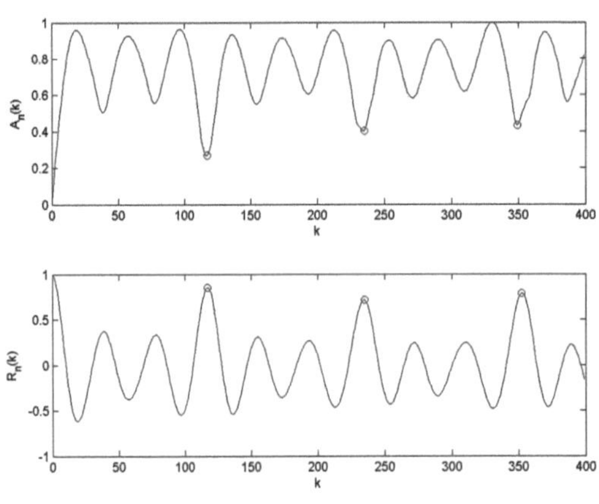

Fig. 7.4. AMDF and auto-correlation functions of an audio frame.

In frequency-based approaches, pitch is determined from the periodic structure in the magnitude of the Fourier transform or cepstral coefficients of a frame. For example, we can determine the pitch by finding the maximum common divider for all the local peaks in the magnitude spectrum. When the required accuracy is high, a large size Fourier transform needs to be computed, which is time consuming.

Spectral Features

The spectrum of an audio frame refers to the magnitude value of the Fourier transform of the samples in the frame. Normally, we use the spectrogram, which is the spectrum of successive overlapping frames, to show the spectral structure of an audio stream. Figure 7.5 shows the spectrograms of three audio clips digitized from TV broadcasts. The commercial clip contains male speech over a music background, the news clip includes clean male speech, and the sports clip is from a live broadcast of a basketball game. Obviously, the difference among these three clips is more noticeable in the frequency domain than in the waveform domain. Therefore, features computed from the spectrum are likely to help audio content analysis.

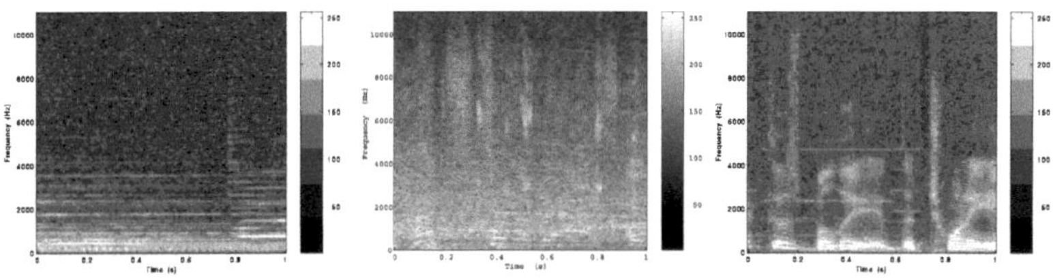

(a) Commercial (b) Basketball (c) News

Fig. 7.5. Spectrums of three audio clips

The difficulty with using the spectrum itself as a frame-level feature lies in its very high dimensionality. For practical applications, it is necessary to find a more succinct description. Let $S_n(\omega)$ denote the power spectrum (i.e. squared magnitude of the spectrum) of frame n. If we think of ω as a random variable, and $S_n(\omega)$ normalized by the total power as the probability density function of ω, we can define the mean and standard deviation of ω. It is easy to see that the mean measures the frequency centroid (FC), whereas the standard deviation measures the bandwidth (BW) of the signal. They are defined as

$$FC(n) = \frac{\int_0^\infty \omega S_n(\omega) d\omega}{\int_0^\infty S_n(\omega) d\omega} \tag{7.4}$$

$$BW^2(n) = \frac{\int_0^\infty (\omega - FC(n))^2 S_n(\omega) d\omega}{\int_0^\infty S_n(\omega) d\omega} \tag{7.5}$$

It has been found that FC is related to the human sensation of the brightness of a sound we hear.

In addition to FC and BW, Liu et al. proposed to use the ratio of the energy in a frequency subband to the total energy as a frequency domain feature [Liu98], which is referred to as the energy ratio of the subband (ERSB). Considering the perceptual property of human ears, the entire frequency band is divided into four subbands, each consisting of the same number of critical bands, where the critical bands correspond to cochlear filters in the human auditory model [Rabiner93]. Specifically, when the sampling rate is 22,050 Hz, the frequency ranges for the four subbands are 0–630 Hz, 630–1720 Hz, 1720–4400 Hz and 4400–11,025 Hz. Because the summation of the four ERSBs is always one, only first three ratios were used as audio features, referred as ERSB1, ERSB2, ERSB3, respectively.

Scheirer et al. used a spectral rolloff point as a frequency domain feature [Scheirer97], which is defined as the 95th percentile of the power spectrum. This is useful to distinguish voiced from unvoiced speech. It is a measure of the "skewness" of the spectral shape, with a right-skewed distribution having a higher value.

Mel-frequency cepstral coefficients (MFCC) or cepstral coefficients (CC) [Rabiner93] are widely used for speech recognition and speaker recognition. While both of them provide a smoothed representation of the original spectrum of an audio signal, MFCC further considers the nonlinear property of the human hearing system with respect to different frequencies. Based on the temporal change of MFCC, an audio sequence can be segmented into different segments, so that each segment contains music of the same style, or speech from one person. Boreczky and Wilcox used 12 cepstral coefficients along with some color and motion features to segment video sequences [Boreczky98].

Linear Predicative Code (LPC)

The linear predicative code is developed based on the physical model of human speech [Huang01]. While articulating, our lungs push the air through the vocal cords, then through the vocal tract (which may include the nasal cavity), and finally out of the mouths. For voiced sound, the vocal cords vibrate at a certain period, which is called pitch, while for unvoiced sound, the vocal cords remain open. The shape of the vocal tract determines the sounds we make. The mathematical model is shown in Fig. 7.6.

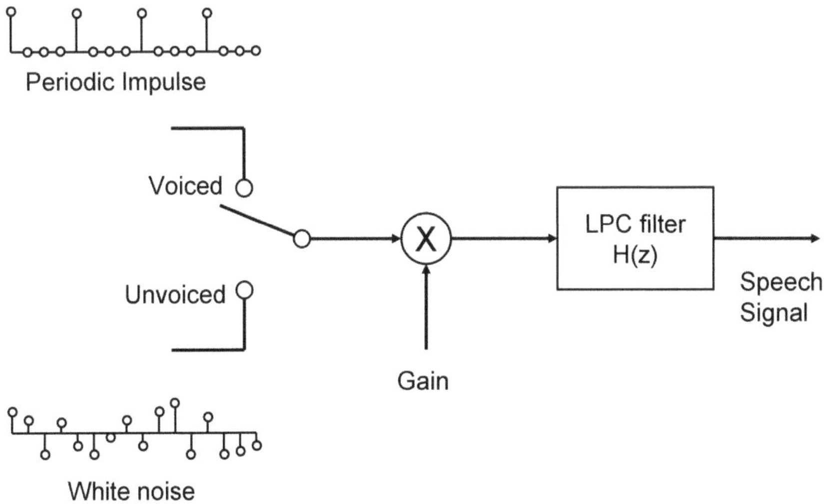

Fig. 7.6. Mathematical model of LPC.

H(*z*) is modeled by an all-pole filter,

$$H(z) = \frac{1}{1 + a_1 z^{-1} + a_2 z^{-2} + \cdots + a_p z^{-p}} = \frac{1}{A(z)} \tag{7.6}$$

It is clear that this model predicts the current sample by a linear combination of its previous *p* samples, and in fact, *H*(*z*) represents a smoothed spectrum of the original speech signal. By minimizing the prediction error, we can determine the LPC coefficients $\{a_1 \ldots a_p\}$ by either the covariance method or the autocorrelation method. There are a few very efficient algorithms for computing the LPC coefficients, e.g. the Levinson–Durbin, lattice, and Schur algorithms [Huang01].

Many acoustic features are derived from or are highly pertinent to LPC coefficients. They include partial correlation (PARCOR) coefficients, Log area ratio (LAR) coefficients, line spectrum pair (LSP) coefficients, line spectrum frequency (LSF) coefficients, etc. [Rabiner93].

7.3.2 Clip-Level Features

As described before, frame-level features are designed to capture the short-term characteristics of an audio signal. To extract the semantic content, we need to observe the temporal variation of frame features on a longer time scale. This consideration leads to the development of various clip-level features, which characterize how frame-level features change over a clip. Therefore, clip-level features can be grouped by the type of frame-level features that they are based-on.

Volume-Based

To measure the variation of volume, Liu et al. proposed several clip-level features [Liu98]. The volume standard deviation (VSTD) is the standard deviation of the volume over a clip, normalized by the maximum volume in the clip. The volume dynamic range (VDR) is defined as $[\max(v) - \min(v)] / \max(v)$, where $\min(v)$ and $\max(v)$ are the minimum and maximum volume within an audio clip. Obviously these two features are correlated, but they do carry some independent information about the scene content. Another feature is volume undulation (VU), which is the accumulation of the difference of neighboring peaks and valleys of the volume contour within a clip.

Scheirer proposed to use the percentage of "low-energy" frames [Scheirer97], which is the proportion of frames with RMS volume less than 50% of the mean volume within one clip. Liu et al. used a non-silence-ratio (NSR) [Liu98], the ratio of the number of non-silent frames to the total number of frames in a clip, where silence detection is based on both volume and ZCR.

The volume contour of a speech waveform typically peaks at 4 Hz. To discriminate speech from music, Scheirer et al. proposed a feature called the 4 Hz modulation energy (4ME) [Scheirer97], which is calculated based on the energy distribution in 40 subbands. Liu et al. proposed a different definition that can be directly computed from the volume contour. Specifically, it is defined as

$$4ME = \frac{\int_0^\infty W(\omega)|C(\omega)|^2 d\omega}{\int_0^\infty |C(\omega)|^2 d\omega} \tag{7.7}$$

where $C(\omega)$ is the Fourier transform of the volume contour of a given clip and $W(\omega)$ is a triangular window function centered at 4 Hz. Speech clips usually have higher values of 4ME than music or noise clips.

ZCR-Based

Normally for speech signals, low and high ZCR segments are interlaced. This is because voiced and unvoiced sounds often occur alternatively in a speech. This is a distinctive characteristic of speech from other types of audio, including music or stadium noise background. Liu et al. used the standard deviation of ZCR (ZSTD) within a clip to classify different audio contents [Liu98]. Saunders proposed to use four statistics of the ZCR as features [Saunders96]. These are: (1) standard deviation of first order difference; (2) third central moment about the mean; (3) total number of zero crossings exceeding a threshold; and (4) difference between the number of zero crossings above and below the mean values. Combined with the volume information, the proposed algorithm can discriminate speech and music at a high accuracy of 98%.

Pitch-Based

The patterns of pitch tracks of different audio contents vary a lot. For speech clips, voiced segments have smoothly changing pitch values, while no pitch information is detected in silent or unvoiced segments. For audio with prominent noisy background, no pitch information is detected either. For a gentle music clip, since there are always dominant tones within a short period of time, many of the pitch tracks are flat with constant values. The pitch frequency in a speech signal is primarily influenced by the speaker (male or female), whereas the pitch of a music signal is dominated by the strongest note that is being played. It is not easy to derive the scene content directly from the pitch level of isolated frames; but the dynamics of the pitch contour over successive frames appear to reveal the scene content more.

Three clip-level features can be used to capture the variation of pitch [Liu98]: standard deviation of pitch (PSTD), smooth pitch ratio (SPR), and non-pitch ratio (NPR). SPR is the percentage of frames in a clip that have similar pitch as the previous frames. This feature is used to measure the percentage of voiced or music frames within a clip, since only voiced and music have smooth pitch. On the other hand, NPR is the percentage of frames without pitch. This feature can measure how many frames are unvoiced speech or noise within a clip.

Frequency-Based

Given frame-level features that reflect frequency distribution, such as FC, BW, and ERSB, one can compute their mean values over a clip to derive corresponding clip-level features. Since the frames with a high energy have more influence on the perceived sound by the human ear, Liu et al. proposed using a weighted average of corresponding frame-level features, where the weighting for a frame is proportional to the energy of the frame. This is especially useful when there are many silent frames in a clip because the frequency features in silent frames are almost random. By using energy-based weighting, their detrimental effects can be removed.

Zhang and Kuo used spectral peak tracks (SPTs) in a spectrogram to classify audio signals [Zhang99]. First, SPT is used to detect music segments. If there are tracks which stay at about the same frequency level for a certain period of time, this period is considered a music segment. Then, SPT is used to further classify music segments into three subclasses: song, speech with music, and environmental sound with music background. Song segments have one of three features: ripple-shaped harmonic peak tracks due to voice sound, tracks with longer duration than speech, and tracks with fundamental frequency higher than 300 Hz. Speech with music background segments have SPTs concentrated in the lower to middle frequency bands and have lengths within a certain range. Those segments without certain characteristics are classified as environmental sound with music background.

There are many other useful audio clip features. Interested readers are referred to [Chang96, Lienhart99, Minami98].

7.4 Audio Segmentation

Audio segmentation is the task of finding the abrupt changes along the audio stream. This task is domain specific, and needs different approaches for different requirements. In this section, we present two segmentation tasks we investigated at two different levels. One is to segment speaker boundaries at the frame level, and the other is to segment audio scenes, for example, commercials and news reporting in broadcast programs at the clip level.

7.4.1 Speaker Segmentation

Speaker segmentation is important for speech recognition and audio content analysis. With speaker adaptation, speech recognizers can significantly improve the accuracy. Speaker information also provides useful cues for indexing audio content. For example, a video conference call may have a few participants, and it is a very useful feature to browse the content of certain speakers. More information about speaker segmentation can be found in [Tranter06].

Liu and Saraclar proposed an iterative speaker segmentation method in [Liu07]. The audio is first segmented into short segments on the phoneme level, where the duration of each segment is in the range of 200 ms to 1 second. Similar to other tasks such as speaker gender classification, the same 39 MFCC features are adopted and each speaker is modeled using a GMM model. Bayesian Information Criteria (BIC) [Chen98] is employed to measure how the speaker models fit the data. Figure 7.7 shows the diagram of the developed algorithm.

The iterative speaker segmentation contains a loop of speaker splitting procedures. In each iteration, the algorithm first increases the number of speakers (NS), and then evaluates all possible splits: splitting speaker i (i = 1, ..., NS–1) into speakers i and NS. For each possible split, the speaker split procedure is applied, and the corresponding BIC value (BIC_{NS}) and speaker labels (L_{NS}) are computed. Among the NS–1 ways of splitting, the one with the maximum BIC value is chosen, and its BIC value and speaker labels are kept as the overall BIC value (BIC_{NS}) and speaker labels (L_{NS}) for the current iteration. The iteration terminates when the BIC value no longer increases.

The speaker label refinement is also an iterative procedure. For each iteration, a set of GMMs is built for all speakers based on the current segment labels. Then all segments are relabeled using the maximum likelihood method based on current speaker models. If the speaker labels converge or the number of iterations reaches a preset value, the refinement iteration stops. Otherwise, a new iteration starts.

The post-processing step merges adjacent segments with the same speaker labels, and smoothes the segments that are too short, e.g. less than 300 ms. Short segments are merged into the longer neighboring segments.

The presented algorithm also detects the change of acoustic channel properties, for example, if the same speaker moved to a different environment, a segment boundary will be declared, although it is not a real speaker change. Testing on two half hour news sequences, 93% true speaker boundaries are detected with a false alarm rate of 22%.

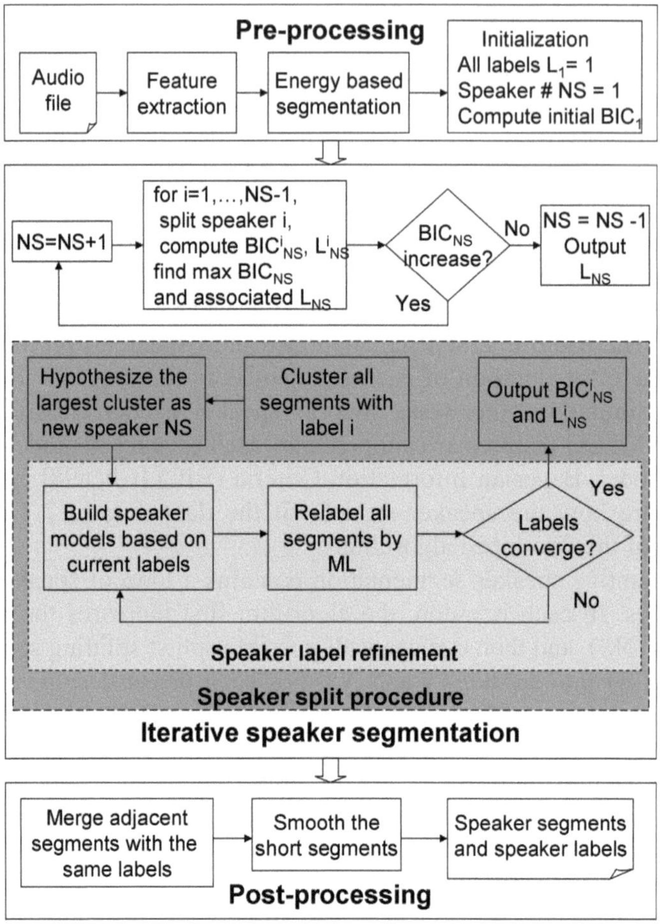

Fig. 7.7. Speaker segmentation algorithm.

7.4.2 Audio Scene Segmentation

Here, the audio scenes we considered are different types of TV programs, including news reporting, commercial, basketball, football, and weather forecasts. To detect audio scene boundaries, a 14 dimensional audio feature vector is computed over each audio clip. The audio features consist of VSTD, VCR, VU, ZSTD, NSR, 4ME, PSTD, SPR, NPR, FC, BW, ERSB1, ERSB2, ERSB3. For a clip to be declared as a scene change, it must be similar to all the neighboring future clips, and different from all the neighboring previous clips. Based on this criterion, we propose using the following measure:

$$\textit{Scene-change-index} = \frac{\left\| \frac{1}{N}\sum_{i=-N}^{-1} f(i) - \frac{1}{N}\sum_{i=0}^{N-1} f(i) \right\|^2}{\sqrt{(c+\mathrm{var}(f(-N),\ldots,f(-1)))(c+\mathrm{var}(f(0),\ldots,f(N-1)))}}, \tag{7.8}$$

where $f(i)$ is the feature vector of the i-th clip, with $i = 0$ representing the current clip, $i > 0$ a future clip, and $i < 0$ a previous clip, $\| * \|$ is the L2 norm, var(…) is the average of the squared Euclidean distances between each feature vector and the mean vector of the N clips considered, and c is a small constant to prevent division by zero. When the feature vectors are similar within the previous N clips and the following N clips, respectively, but differ significantly between the two groups, a scene break is declared. If two breaks are closer than N clips away, the one with smaller scene-change-index value is removed. The selection of the window length N is critical: If N is too large, this strategy may fail to detect scene changes between short audio shots. It will also add unnecessary delay to the processing. Through trial-and-error, it is found that $N = 6$ gives satisfactory results [Liu98].

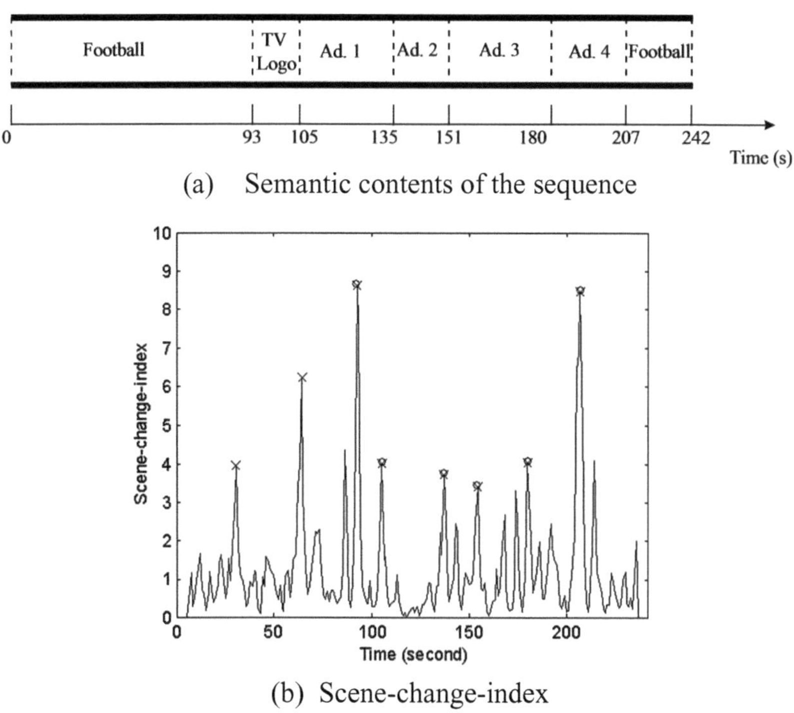

Fig. 7.8. Content and scene-change-index for one audio stream.

Figure 7.8(a) shows the content of one testing audio sequence used in segmentation. This sequence is digitized from a TV program that contains seven different semantic segments. The first and the last segments are both football games, between which are the TV station's logo shot and four different commercials. The duration of each segment is also shown in the graph. Figure 7.8(b) shows the scene-change-index computed for this sequence. Scene changes are detected by identifying those clips for which the scene-change-indices are higher than a threshold, D_{min}. D_{min} is set to 3, which have been found to yield good results through trial-and-error. In these graphs, mark "o" indicates real scene changes and "*" detected scene changes. All the real scene changes are detected using this algorithm. Note that there are two falsely detected scene changes in the first segment of the sequence. They correspond to the sudden appearance of the commentator's voice and the audience's cheering.

7.5 Audio Content Categorization

After audio segmentation, we need to classify each segment into predefined categories. The categories are normally semantically meaningful high-level labels that are determined from low-level features. For example, for speech, the categories can be spoken languages, speaker genders, speaker identifications, etc. For music, the categories can be classic, pop, jazz, and so on. The pattern recognition mechanism fits in this gap, and maps the distribution of low level features to high level semantic concepts. In the section, we will present three different audio classification situations: speaker recognition, speech/non-speech classification, and music genre classification.

7.5.1 Speaker Recognition

The goal of speaker recognition is to automatically recognize the identification of a speaker based on his/her voice. Speaker recognition provides complementary information for other biometric authentication methods, including fingerprints, face, and iris. It has wide applications, including user authentication in a dialog system, surveillance, audio indexing and retrieval, and forensic speaker recognition. For example, occurrences of the anchorpersons in broadcast news often indicate semantically meaningful boundaries for reported news stories. Researchers have been working on this topic for more than three decades. To push forward more effective approaches, NIST has been coordinating speaker recognition evaluation since 1996.

There are two main tasks in speaker recognition: speaker verification and speaker identification. The speaker verification task is to verify the claimed identity of a speaker. This is refereed to as an open-set task since the imposters are not known to the system. Speaker identification refers to the task of identifying a speaker from a set of known speakers. This is refereed to a closed-set task since the input voice is from a prior known speaker.

Depending on whether the spoken text is predefined or not, speaker recognition contains two operation modes: text-dependent and text-independent. For the text-dependent case, the speaker is prompted to repeat certain text, either a combination of numbers or phrases. Since the system knows what the speaker speaks, more effective algorithms can be applied to achieve higher recognition accuracy. For the text-independent case, the speaker can speak anything they want. Surely, this is a more flexible mode of operation, but the system will need more data for training, and the overall performance is not as good as the text-dependent cases.

Many acoustic features have been investigated for speaker recognition tasks in various applications with different constraints and requirements. Acoustic features for speaker recognition should have high speaker discrimination power, which means high inter-speaker variability and low intra-speaker variability. Adopted features include linear prediction coefficients (LPC), filter-bank, cepstrum coefficients, log-area ratio (LAR), LSP, MFCC, etc. within which MFCC gains more prevalence due to its effectiveness [Campbell97]. In terms of speaker pattern matching methods, there are generally two categories of approaches: template model and statistical model. Vector quantization, nearest neighbor, and dynamic time warping (DTW) based on certain distance measures (e.g. the Mahalanobis distance) belong to the first category. Gaussian mixture models, hidden Markov models, and support vector machines are the most popular ones in the second category.

The performance of speaker recognition systems varies in a wide range since the amount and the quality of the training/testing data are very different [Faundez05]. Reynolds [Reynolds02] provided an overview of the performance achieved by four typical systems shown in Table 7.1. The equal error rate (ERR) points are used as a performance summary, where ERR is an indicator where the false alarm is equal to the false recognition.

Huang et al. studied anchorperson detection, which can be categorized as a speaker verification problem [Huang99]. Detection of anchorperson segments is carried out using text independent speaker recognition techniques. The target speaker (anchorperson) and background speakers are represented by 64 component Gaussian mixture model with diagonal covariance matrices. Again, the audio features utilized are 13 MFCC coefficients and their first and second order derivatives to form 39 features in total. A maximum likelihood classifier is applied to detect the target speaker segments. Testing on a dataset of four half-hour news sequences, this approach successfully detects 91.3% of real anchorperson speech, and the false alarm rate is 1%.

Table 7.1. Performance of typical speaker verification systems.

Task	Text-dependent		Text-independent	
Data type	Combinations lock phrases	10 digit string	Conversational speech	Read sentences
Data quality	Clean telephone data			Noisy Radio data
Enrollment	3 minutes	2 strings	2 minutes	30 seconds
Testing	2 seconds	1 string	30 seconds	15 seconds
ERR (%)	0.1 - 1	1-5	7-15	20-35

While most classifiers described in this section, e.g. GMM, try to model the feature density of each speaker, it is interesting to introduce the concept of discriminative learning, where the focus is on learning the class boundaries. Studies have shown that discriminative learning can further improve the speaker identification performance.

7.5.2 Audio Scene Detection

Audio scenes are segments with homogeneous content in an audio stream. For example, broadcast news programs generally consist of two different audio scenes: news reporting and commercials. Discriminating them is very useful for indexing news content. One obvious usage is to create a summary of news program, where commercials segments are removed.

Depending on the application, different categories of audio scenes and different approaches are adopted. Saunders [Saunders96] considered the discrimination of speech from music. Saraceno and Leondardi further classified audio into four groups: silence, speech, music, and noise [Saraceno97]. The addition of the silence and noise categories is appropriate, since a large silence interval can be used as segment boundaries, and the characteristic of noise is very different from that of speech or music.

A more elaborate audio content categorization was proposed by Wold et al. [Wold96], which divides audio content into ten groups: animal, bells, crowds, laughter, machine, instrument, male speech, female speech, telephone, and water. To characterize the difference among these audio groups, Wold et al. used mean, variance, and auto correlation of loudness, pitch, brightness (i.e. frequency centroid) and bandwidth as audio features. A nearest neighbor classifier based on a weighed Euclidean distance measure was employed. The classification accuracy is about 81% over an audio database with 400 sound files.

Video search can obviously benefit from the ability to classify media based on the audio content. For example, TV broadcasts can be classified into categories such as news reporting, commercial, weather forecast, basketball game, and football game [Liu98]. Based on a set of 14 audio features extracted from audio energy, zero crossing rate, pitch, and spectrogram, a three layer feed forward neural network classifier achieves 72.5% accuracy. A classifier based on hidden Markov model further increases the accuracy by 12%.

Another interesting study related to general audio content classification is by Zhang and Kuo [Zhang99]. They explored five kinds of audio features: energy, ZCR, fundamental frequency, timber, and rhythm. Based on these features, a hierarchical system for audio classification and retrieval was built. In the first step, audio data is classified into speech, music, environmental sounds, and silence using a rule-based heuristic procedure. In the second step, environmental sounds are further classified into applause, rain, bird sound, etc. using an HMM classifier. These two steps provide the so-called coarse-level and fine-level classification. The coarse-level classification achieves 90% accuracy and the fine-level classification achieves 80% accuracy in a test involving 10 sound classes.

7.5.3 Music Genre Classification

Digital music, in all kinds of formats including MPEG Layer 3 (MP3), Microsoft Windows Media format, RealAudio, MIDI, etc. is a very popular type of traffic in the Internet. When music pieces are created, they are normally assigned with related metadata by producers or distributors, for example, title, music category, author name, and date. Unfortunately, most of the metadata is not available or is lost in the stages of music manipulation and format conversion. Music genre, as a specific metadata, is important and indispensable for music archiving and querying. For example, a simple query to find all pop music in a digital music database requires the category information. Since manually re-labelling is time consuming and inconsistent, we need an automatic way to classify music genre.

Music genre classification has attracted a lot of research effort in recent years [Slaney08]. Tzanetakis et al. [Tzanetakis02] explored the automatic classification of audio signals into a hierarchy of music genres. On the first level, there are 10 categories: classical, country, disco, hiphop, jazz, rock, blues, reggae, pop, and metal. On the second level, classical music is further separated into choir, orchestra, piano, and string quartet, and jazz is further split into bigband, cool, fusion, piano, quartet, and swing. Three sets of audio features are proposed, which reflect the timbral texture, rhythmic content and pitch content of audio signal, respectively. Timbral texture features include spectral centroid, spectral rolloff, spectral flux, zero crossing rate, and MFCC. Rhythmic content features are calculated based on the wavelet transform, where the information of main beat, subbeats and their periods and strengths are extracted. Pitch content features are extracted based on multiple pitch detection techniques. Utilized pitch features include the amplitude and period of the maximum peaks of the pitch histogram, pitch interval between the two most prominent peaks of the pitch histogram, and the sum of the histogram. Tzanetakis et al. tested different classifiers, including a simple Gaussian classifier, a Gaussian mixture model classifier, and a *K*-nearest neighbour classifier. Among them, GMM with 4 mixtures achieves the best classification accuracy, which is 61%. Considering that a human being makes 20–30% errors on classifying musical genre in a similar task, the performance of automatic music genre classification is reasonably good.

Lambrou et al. [Lambrou98] investigated the task of classifying an audio signal into three different music styles: rock, piano, and jazz. They used zero crossing rate and statistical signal features in the wavelet transform domain as acoustic features. Overall, seven statistics are computed including first order statistics: mean, variance, skewness, and kurtosis, and second order statistics: angular second moment, correlation, and entropy. Lambrou et al. benchmarked four different classifiers: minimum distance classifier, *K*-nearest neighbors classifier, least squares minimum distance classifier (LSMDC), and quadrature classifier. Simulation results show that LSMDC gives the best performance with an accuracy of 91.67%.

7.6 Speech Recognition

Automatic speech recognition (ASR) is a process to convert a stream of acoustic signals into a sequence of words by machines. Built on decades of intensive research and engineering, ASR technology has reached its maturity for a wide range of real applications, from speaker dependent dictation tasks to speaker independent very large vocabulary conversational speech recognition tasks. Lexicons and grammars may range from simple command and control and digit recognition to 200,000 word vocabularies or even systems for recognizer one million names. Systems are developed to handle acoustic conditions from mobile telephony up through broadcast quality speech.

Most state of the art speech recognizers adopt a statistical approach [Jelinek97]. Let **A** denote a sequence of acoustic features, $\mathbf{A} = \{a_1, a_2, \ldots a_i, \ldots, a_I\}$, where each a_i is a feature vector, and let **W** denote a string of words, $\mathbf{W} = \{w_1, w_2, \ldots, w_j, \ldots, w_J\}$, where each w_i is a word. The core task of speech recognition is to find the word string **W'**, that maximize the probability of $p(\mathbf{W}|\mathbf{A})$.

Following the Bayes' formula, $p(\mathbf{W}|\mathbf{A})$ can be rewritten as,

$$p(\mathbf{W} \mid \mathbf{A}) = \frac{p(\mathbf{W})p(\mathbf{A} \mid \mathbf{W})}{p(\mathbf{A})} \quad (7.9)$$

Since $p(\mathbf{A})$ is a constant that does not depend on **W**, it can be ignored while searching **W'** as follows,

$$\mathbf{W}' = \arg\max_{\mathbf{W}} p(\mathbf{W})p(\mathbf{A} \mid \mathbf{W}) \quad (7.10)$$

$p(\mathbf{A}|\mathbf{W})$ is the probability of speech feature **A** while **W** is spoken, and it relies on the lexicon, which determines the pronunciations of words, and the acoustic model, which models the sound units (e.g. phonemes) based on speech features. $p(\mathbf{W})$ is the a priori probability of **W**, which can be decomposed as,

$$p(\mathbf{W}) = \prod_{j=1}^{J} p(w_j \mid w_1, \ldots w_{j-1}) \approx \prod_{j=1}^{J} p(w_j \mid w_{j-n+1}, \ldots w_{j-1}) \quad (7.11)$$

The approximation in the formula is a practical way to reduce the complexity of language model. This kind of language model is called an *n-gram*, where the probability of speaking word w_j only depends on the last $n-1$ words. A popular choice in many real speech recognition systems is 3-gram.

Figure 7.9 shows the block diagram of a typical ASR system. It is composed of two major components: the front end and the decoder. The front end block extracts spectrum representation of the speech waveform. The most widely used features are Mel Frequency Cepstral Coefficients (MFCC). The decoder block searches the best match of word sequences for the input acoustic features based on acoustic model, lexicon, and language model.

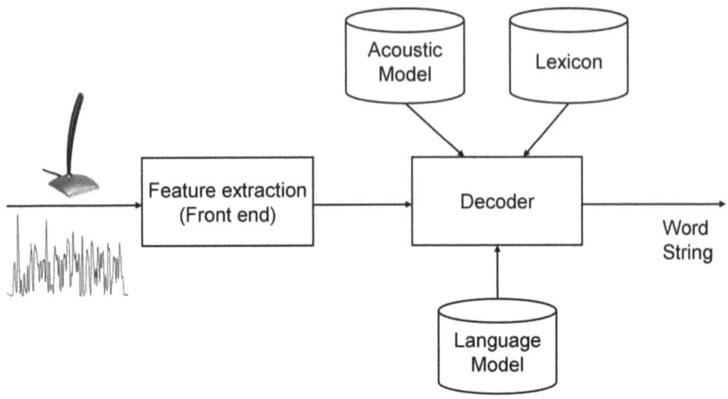

Fig. 7.9. Diagram of automatic speech recognition system.

Speech recognition is still an active research area in many academic and industrial laboratories, including CMU, MIT, Microsoft, AT&T, BBN, IBM, Nuance, etc. AT&T has been well known for its contributions in speech recognition. In the following we briefly describe the architecture of the Watson speech recognition engine developed at AT&T Research Labs.

AT&T Watson [Goffin05] is a real-time low latency speech recognizer, which utilizes continuous-density hidden Markov models for acoustic modeling and finite state networks for language modeling. The core recognizer of Watson is built on a Controller (CTL), which has access to a Data Store (DS) and an Execution Context (EC). The DS handles all data used by the recognizer, including word dictionaries, acoustic models, and language models. The EC represents the algorithm pipeline, including feature extraction and normalization, endpointing, barge-in detection, decoding, and scoring.

Figure 7.10 shows a real time, low latency, large vocabulary speech recognition system for broadcast news developed by AT&T. The vocabulary is over 210,000 words, and the achieved word accuracy is typically between 75 and 95% depending on the conditions.

7.7 Audio Query and Browsing Techniques

Audio content query and browsing is an as important issue for content analysis. In this section, we first present SpeechLogger [Saraclar04b], a research system for searching and browsing spoken documents, or the spoken component of multimedia communications, and then show an audio query by example system.

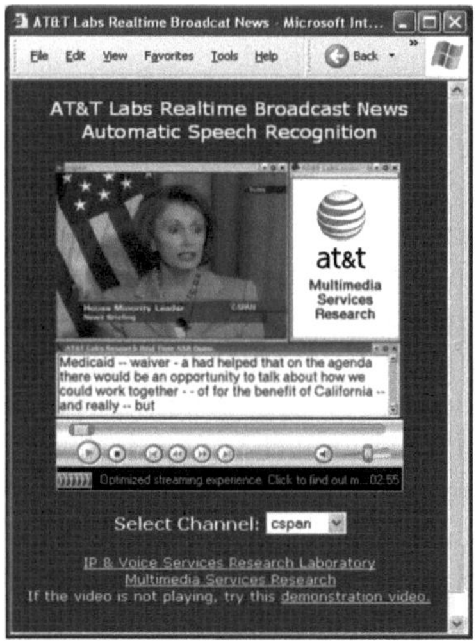

Fig. 7.10. AT&T real time broadcast news speech recognition system.

7.7.1 SpeechLogger

Figure 7.11 shows the overview of SpeechLogger system. The audio can be recorded via telephone or microphone or can be prerecorded. Alternatively, the audio can be obtained by separating the audio from video broadcasts. Once various speech processing techniques are applied and the speech is indexed, it is possible to search and browse the audio content.

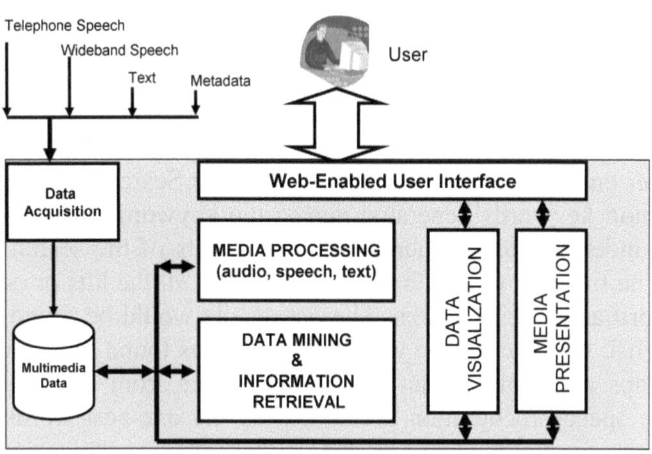

Fig. 7.11. SpeechLogger system.

Once a new completed audio recording is available, the following audio processing steps can begin: speaker segmentation, speech recognition, transcription alignment, keyword extraction, audio compression, and speech indexing. Each step will be described in more detail below. We attempt to distinguish the different speakers from each other in the speaker segmentation component. The speech recognition component includes various approaches including best word, best phone, word lattice, phone lattice, and hybrid hypotheses. If a transcript is available, the transcript can be synchronized (or aligned) in time with the speech recognition output. The keyword extraction component generates the most salient words found in the speech recognition output (best word) or transcript (if available) and can be used to determine the nature of the spoken communications. The audio compression component compresses the audio file and creates an MP3 audio file which is copied to a media server for delivery and presentation via the Web-enabled user interface. The final step in the processing is text and lattice indexing.

The user interface applies for telephone conversations, teleconferences, and broadcast news, although the audio and speaker quality does vary for each of these types of spoken communications. Here, we focus on teleconference call recording as a use scenario. Once the user has found the desired call using one of the retrieval modules (one-best word, one-best phone string, word lattice, phone lattice, or both word and phone lattice), the user can navigate the call using the user interface elements described below.

One-Best Word Search

For the one-best word index, Fig. 7.12 shows the user interface for searching, browsing, and playing back spoken documents. The user can browse the call at any time by clicking on the timeline to start playing at that location on the timeline. The compressed audio file (MP3) that was created during the processing would be streamed to the user. The user can at any time either enter a word (or word phrase) in the Search box or use one of the common keywords generated during the keyword extraction process. The text index would be queried and the results of the search would be shown. The timeline plot at the top would show all the hits or occurrences of the word as thin tick marks. The list of hits would be found under the keyword list. In this case, the word "chapter" was found four times and the time stamps are shown. The time stamps come from the results of the automatic speech recognition process when the one-best words and time stamps were generated. The search term "chapter" is shown in bold with five context words on either side. The user can click on any of these four hits to start playing where the hit occurred. The solid band in the timeline indicates the current position of the audio being played back. The entire call, in this case, is 9:59 minutes long and the audio is playing at the beginning of the fourth hit at 5:20 minutes. As part of the processing, caption data is generated in Microsoft's SAMI (Synchronized Accessible Media Interchange) format from the one-best word output in order to show caption text during the playback. The caption text under the timeline will be updated as the audio is played. At this point in the call, the caption text is "but i did any chapter in a". This caption option can be disabled by clicking on the CC icon and can be enabled by clicking on the CC icon again. The user can also speed up or slow down the playback at any time by using the "Speed" button. The speed will toggle from 50% (slow) to 100% to 150% (fast) to 200% (faster) and then start over at 50%. This allows the user to more quickly peruse the audio file. Techniques such as the Waveform Similarity Overlap-and-Add (WSOLA) [Verhelst93] can be used to increase playback rate while preserving pitch, and pause removal can be employed to increase the intelligibility of speech playback at higher speed.

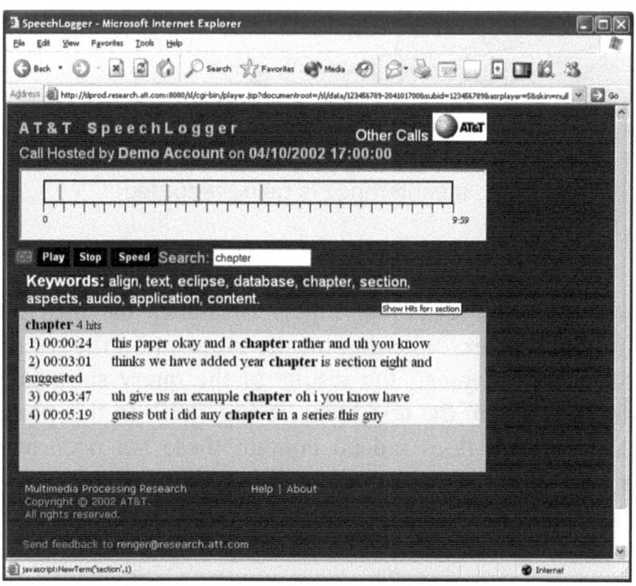

Fig. 7.12. User interface for ASR text search.

Lattice and Phonetic Search

Figure 7.13 shows a word lattice (highly-pruned) that is used in the recognition process [Saraclar04a]. Each arc corresponds to a word. The normalized probability of each world is shown next to the word. The 1-best transcript approach finds the path inside the lattice with the maximum probability. In this example this is the path that represents the string of words “a conference is being recorded.” While this is mostly correct, the correct string of words in the audio is “our conference is being recorded.” By retaining only the 1-best transcription, we lose the possibility of finding any hits for the word “our” (which actually is in the spoken words). Searching the lattice instead of the 1-best transcript, however, allows for this word to be found.

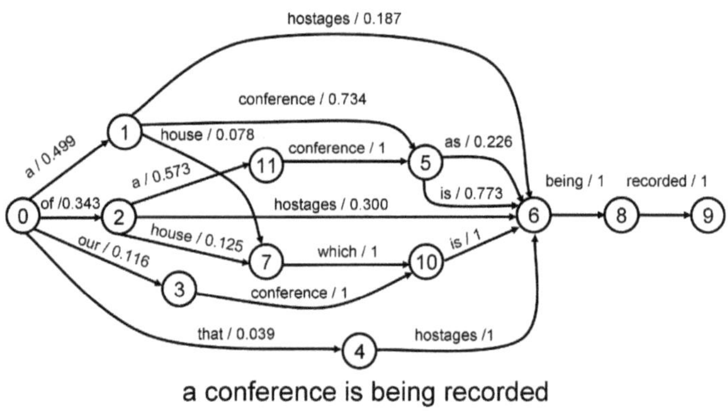

Fig. 7.13. Lattice-based speech search.

A similar Web application in Fig. 7.14 shows the user interface for searching a lattice index. Note that for the same audio file (or call) and the same search term “chapter”, the results of the query show six hits compared to the four hits in the text index in Fig. 7.12. In this particular case, the manual transcript does indeed contain these six occurrences of the word “chapter.” The search terms were found in audio segments, which is why the time of the hit is a time range. The information in brackets is the expected count and can exceed 1.0 if the search term occurs more than once in the audio segment. The time range is reflected in the timeline where the thin tick marks have been replaced with colored segments. The colors of the segments correspond to the colors of the hits in the list. The darker the color, the higher the count and the lighter the color, the lower the count. Finally, the search can be refined by altering the threshold using the “Better Hits” and “More Hits” buttons. In this example, the threshold is set to 0.2 as can be seen under the timeline. If the user clicks on the “Better

Hits" button, the threshold is increased so that only more reliable matches are shown. If the "More Hits" button is used, the threshold is decreased so more hits are shown although the hits may not be as reliable (i.e. they may include false positives). The lattice index only returns hits where each hit has a count above the threshold.

The lattice search user interface allows the user to more easily find what the user wants and has additional controls (threshold adjustments) and visual feedback (colored segments/hits) that are not possible with the 1-best text search user interface.

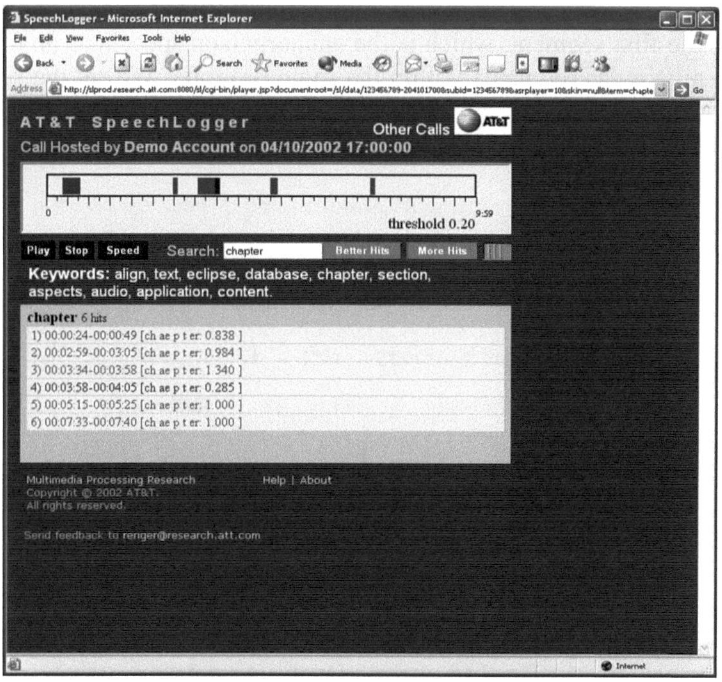

Fig. 7.14. User Interface for lattice search.

7.7.2 Query by Example

The overwhelming majority of Web queries are textual. With the fast development of multimedia applications, not only has the demand outgrown the capabilites of textual queries, but also manual annotation is no longer feasible. Query based on acoustic characteristics is one alternative to text based retrieval. For example, to retrieve audio clips based on text, one has to know exactly how this clip is labelled. But there are many cases where users know only what the content (speakers, music, or songs) sounds like but not the semantics with which it has been identified previously. Therefore, retrieval by audio example is an alternative to conventional text based retrieval. The user can simply provide a sample audio stream and ask to retrieve the audio segments that possess similar acoustic properties.

In this section, we show a simple example of audio query by example [Liu00]. Figure 7.15 shows the user interface. The audio data is first segmented such that each segment contains one speaker. Each segment is then fitted by a GMM and then a distance matrix is computed to measure the difference between any pair of audio segments. In the example shown, we use the first segment, which is the anchorperson speech, as an example to find all similar segments within the same program. We set the sensitivity to 0.55 and find 14 segments, all of which are actually the same anchorperson.

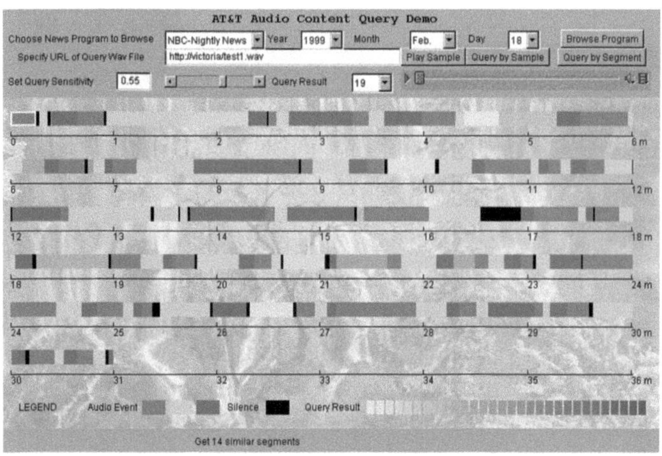

Fig. 7.15. Audio query by example.

7.8 Conclusion

This chapter reviewed the recent progress of audio content processing. Audio content indexing and retrieval plays an important role in the field of multimedia information retrieval. Here, we outline several possible future directions in this field to conclude this chapter. First, it remains a challenge to find an application independent approach in unrestricted domains. Audio content analysis in specific domains has been intensively studied, and researchers have found feasible solutions to most of the problems. Although different applications share many low-level audio processing technologies, there is not a general framework that universally fits most applications. We believe that this topic deserves more research effort, since the range of applications is growing rapidly with more and more audio retrieval systems in various domains being developed. Second, personalization demands more attention in future audio retrieval systems. Most of the currently available audio retrieval systems ignore the individual characteristics of end users. Personalized profiles for each user, which log the history of user queries and activities, user preferences, etc. are useful for the audio retrieval system to generate better query results tuned for a particular user. Third, commercialization is possible with the current state of audio analysis capability. Because of the research effort invested in recent years, audio retrieval systems in certain fields achieve acceptable performance, and they provide substantial benefit for users to find audio content fast and easily. Computational complexity of audio indexing and query continues to be a bottleneck. More efficient technologies are indispensable to increase system scalability and reduce infrastructure cost for commercial systems.

References

[Baluja07] Baluja, S. and Covell, M.: Audio fingerprinting: combining computer vision & data stream processing. *ICASSP* (2007).

[Boreczky98] Boreczky, J. S. and Wilcox, L. D.: A hidden Markov model framework for video segmentation using audio and image features. *ICASSP*, **6**, pp. 3741–3744 (1998).

[Campbell97] Campbell, J. P. JR.: Speaker recognition: A tutorial. *Proceedings of the IEEE*, **85**(9) (1997).

[Chang96] Chang, Y., Zeng, W., Kamel, I., Alonso, R.: Integrated image and speech analysis for content-based video indexing. *The IEEE International Conference on Multimedia Computing and Systems*, pp. 306–313 (1996).

[Chang01] Chang, S., Sikora, T., Puri, A.: Overview of the MPEG-7 standard. *IEEE Transaction On Circuits and Systems*, **11**(6), pp. 688-695 (2001).

[Chen98] Chen, S. and Gopalakrishnan, P.: Speaker, environment and channel change detection and clustering via the Bayesian information criterion. DARPA Speech Recognition Workshop (1998).

[Faundez05] Faundez-Zanuy, M. and Monte-Moreno, E.: State-of-the-art in speaker recognition. *IEEE Aerospace and Electronics Systems Magazine*, **20**(5), pp. 7–12 (2005).

[Goffin05] Goffin, V., Allauzen, C., Bocchieri, E., Hakkani-Tur, D., Ljolje, A., Parthasarathy, S., Rahim, M., Riccardi, G., Saraclar, M.: The

AT&T Watson speech recognizer. *ICASSP*, **1**, pp. 1033–1036 (2005).

[Hess83] Hess, W.: *Pitch Determination of Speech Signals*. Springer-Verlag (1983).

[Huang99] Huang, Q., Liu, Z., Rosenberg, A.: Automated semantic structure reconstruction and representation generation for broadcast news. *Proc. Of SPIE* (1999).

[Huang01] Huang, X., Acero, A., Hon, X.: *Spoken Language Processing*. Prentice Hall (2001).

[Jelinek97] Jelinek, F.: *Statistical Methods for Speech Recognition*. The MIT Press (1997).

[Kim06] Kim, H., Moreau, N. and Sikora, T.: *MPEG-7 Audio and Beyong: Audio Content Indexing and Retrieval*. Wiley (2006).

[Lambrou98] Lambrou, T., Kudumakis, P., Speller, R., Sandler, M., and Linney, A.: Classification of audio signals using statistical features on Time and wavelet transform domains. *ICASSP*, **6**, pp. 3621–3624 (1998).

[Lienhart99] Lienhart, R., Pfeiffer, S., Effelsberg, W.: Scene determination based on video and audio features. *IEEE International Conference on Multimedia Computing and Systems*, **1**, pp. 685–690 (1999).

[Liu00] Liu, Z. and Huang, Q.: Content-based indexing and retrieval-by-example in audio. *ICME* (2000).

[Liu07] Liu, Z. and Saraclar, M.: Speaker segmentation and adaptation for speech recognition on multiple-speaker audio conference data. *ICME* (2007).

[Liu98] Liu, Z., Wang, Y., Chen, T.: Audio feature extraction and analysis for scene segmentation and classification. J. *VLSI Signal Processing Sys. Signal, Image, Video Technol*, **20**, pp. 61–79 (1998).

[Lu08] Lu, L. and Hanjalic, A.: Audio keywords discovery for text-like audio content analysis and retrieval, *IEEE Transaction On Multimedia*, **10**(1), pp. 74–85 (2008).

[McKin03] McKinney, M., Breebaart, J.: Features for audio and music classification. *ISMIR* (2003).

[Minami98] Minami, K., Akutsu, A., Hamada, H., Tonomura, Y.: Video handling with music and speech detection. *IEEE Multimedia Magazine*, **5**, pp. 17–25 (1998).

[Pfeiffer96] Pfeiffer, S., Fischer, S., Effelsberg, W.: Automatic audio content analysis. *Proc. 4th ACM Int. Conf. Multimedia*, pp. 21–30 (1996).

[Rabiner93] Rabiner, L. and Juang, B.H.: *Fundamentals of Speech Recognition*. Prentice Hall (1993).

[Reynolds02] Reynolds, D.A.: An overview of automatic speaker recognition technology. *ICASSP*, **4**, pp. 4072–4075 (2002).

[Saraceno97] Saraceno, C. and Leonardi, R.: Audio as a support to scene change detection and characterization of video sequences. *ICASSP,* **4**, pp. 2597–2600 (1997).

[Saraclar04a] Saraclar, M. and Sproat, R.: Lattice-based search for spoken utterance retrieval. *HLT/NAACL* (2004).

[Saraclar04b] Saraclar, M., Begeja, L., Gibbon, D., Liu, Z., Renger, B., Shahraray, B.: A system for searching and browsing spoken communications. *HLT/NAACL Workshop on Interdisciplinary Approaches to Speech Indexing and Retrieval*, (2004).

[Saunders96] Saunders, J.: Real-time discrimination of broadcast speech/music. *ICASSP*, **2**, pp. 993–996 (1996).

[Scheirer97] Scheirer, E. and Slaney, M.: Construction and evaluation of a robust multifeatures speech/music discrimination. *ICASSP*, **2**, pp. 1331–1334 (1997).

[Slaney08] Slaney, M., Ellis, D., Sandler, M., Goto, M., Goodwin, M.: Introduction to the special issue on music information retrieval. *IEEE Transaction On Audio, Speech, Lanuage*, **16**(2), pp. 253–254 (2008).

[Thong03] Thong, J., Blackwell, S., Weikart, C. and Mandviwala., H.: Multimedia content analysis and indexing: evaluation of a distributed and scalable architecture. HP technical report, http://www.hpl.hp.com/techreports/2003/HPL-2003-182.ps, cited 10 Dec 2007.

[Tranter06] Tranter, S.E. and Reynolds, D.A.: An overview of automatic speaker diarization systems. *IEEE Transactions On Audio, Speech, and Language Processing*, **14**(5), (2006).

[Tzanetakis02] Tzanetakis, G. and Cook, P.: Musical genre classification of audio signals. IEEE Transactions On Speech and Audio Processing. **10**(5), pp. 293–302 (2002).

[Verhelst93] Verhelst, W. and Roelands, M.: An overlap-add technique based on waveform similarity (WSOLA) for high quality time-scale modification of speech. *ICASSP*, pp. 554–557 (1993).

[Wold96] Wold, E., Blum, T., Keislar, D., Wheaton, J.: Content-based classification, search, and retrieval of audio. *IEEE Multimedia*, **3**(2), pp. 27–36 (1996).

[Zhang99] Zhang, T. and Kuo, C.: Hierarchical classification of audio data for archiving and retrieving. *ICASSP*, **6**, pp. 3001–3004 (1999).

[Zhang00] Zhang, T. and Kuo, C.: *Content-based audio classification and retrieval for audiovisual data parsing*, Springer, (2000).

8 Text Processing

8.1 Introduction

Text provides crucial cues for understanding content. For example, the closed captions in broadcast television programs and subtitles in DVD movies facilitate video consumption for viewers. When a transcript is not available for certain content, automatic speech recognition can be used to extract linguistic information. Text information is much more concise than corresponding audio or video. The reason is that we need language knowledge to understand text, and the knowledge itself does not need to be embedded in the text data. For example, we only need five characters to express a "plane," but to show a video clip of plane takes millions of bytes. Text streams contain very rich semantic information. How to effectively extract information from text is an important component in video content analysis.

Text processing has been studied in different fields for many years. Computational linguistics and natural language processing [Johnston07] are two areas that have produced many text processing techniques. Main tasks in these areas include parser design, tagger design, information extraction, named entity recognition, language modeling, story summarization, topic segmentation, data mining, machine translation, speech recognition, spoken dialogue system, etc. Readers may find additional information in books devoted to text processing [Allen95, Jurafsky00, Manning00] and information retrieval [Baeza99]. In this chapter, we focus on the text processing techniques that are relevant to text document retrieval. Recently, these classical text retrieval techniques were applied to multimedia content query [Feng03].

Text document retrieval has been extensively studied by the Text REtrieval Conference (TREC) conference [TREC07], co-sponsored by the National Institute of Standards and Technology (NIST) and US Department of Defense. TREC supports research within the information retrieval community by providing the infrastructure necessary for large-scale evaluation of text retrieval methodologies. Topic Detection and Tracking (TDT) [Wayne00] is an integral part of the DARPA Translingual Information Detection, Extraction, and Summarization (TIDES) program. The goal of the TIDES program is to enable English-speaking users to access, correlate, and interpret multilingual sources of real-time information and to share the essence of this information with collaborators.

In this chapter, we will introduce some fundamentals in text processing that are relevant to content analysis, information extraction, and information retrieval. Specifically, we will discuss part of speech tagging, named entity extraction, text capitalization, stemming, term weighting, and document ranking. We will also present a few methods for story segmentation and text summarization.

8.2 Story Segmentation

Broadcast news typically contains a number of unrelated stories. The term story segmentation describes segmenting an input program into topically cohesive stories. Story segmentation can be performed using audio, visual, and textual information, combined together. In this section, we focus on text based story segmentation approaches.

The US government initiated Topic Detection and Tracking (TDT) research in 1996 [Wayne00], when the Defense Advanced Research Projects Agency (DARPA) realized that it needed technology to determine the topical structure of news stream without human intervention. The TDT project began in 1997 with a pilot study that included Carnegie Mellon University, the University of Massachusetts, and Dragon Systems. Since then, the project has continued to the present year with annual technology evaluation cycles. Each cycle begins with a statement of the year's evaluation criterion in the evaluation plan, which is followed by a period of research, an evaluation, and finally a workshop to discuss the findings and research.

This section mainly discusses four different approaches for story segmentation: cue phrases, cosine similarity, dynamic programming, and topic classification.

8.2.1 Cue Phrases

In certain applications, a set of words or phrases can often act as cues, indicating the presence of a nearby story boundary. These words/phrases are referred to as cue phrases. In this section, we use broadcast news programs as an example to show how cue phrases can be used to detect story boundaries.

Given that broadcast news is created for a large audience, and is aired within a limited amount of time, the content is well structured, concise, comprehensive, and easy to follow. Patterns are observed in typical news programs, e.g. introduction of a new story by the anchorperson, followed with detailed coverage from the reporters and interviewees. Anchorpersons normally use cue phrases to start the program, for example, “Good evening,” to pass to reporters at the scene after introduction, for example, “NBC's John Yang at The Pentagon,” and to finish the detailed coverage and summarize the story, for example, “Thank you, John.” These cue phrases can be easily and robustly detected and they provide reliable indicator for story changes.

The cue phrases can be classified into more detailed categories. A two category classification can be (1) begin-cue-phrases in the beginning of news, for example, “Good evening” and (2) miscellaneous cue-phrases in the middle of the programs, such as “Weather forecast is next,” “when we come back,” etc. Merlino et al. [Merlino97] proposed a finer classification for cue phrases. They are

1. I’m ‹person›. For example, “I’m John.”
2. Introductory phrases. For example, “Hello and welcome,” “Thanks for watching,” “We are back here with NBC News in-depth.”
3. Weather related cue phrases. For example, “forecast,” “high pressure,” “hurricane.”
4. Anchor to reporter phrases. For example, “NBC’s Lisa Jones,” “Our story tonight from NBC's Bob Smith.”
5. Reporter to Anchor phrases. For example, “Tom Costello, NBC News, Washington.”
6. Story preview. For example, “When we continue after a break here.”

With natural language understanding tools, these cue phrases can be detected by regular expression matching or more data driven approaches, including hidden Markov models.

8.2.2 Cosine Similarity

In text processing, a document can be represented as a vector in a space, where each dimension corresponds to a distinct term in a predefined vocabulary. The coefficients of the vector for a given document are the term frequencies within that dimension. The resulting vectors are extremely sparse and typically high frequency words (stop words) are ignored. Such a representation of text document is called a vector space model.

The cosine coefficient is a document similarity metric which has been investigated extensively. The cosine of the angle between two vectors $\vec{A}$ and $\vec{B}$, each representing a document, is an indication of vector similarity and is equal to the dot product of the vectors normalized by the product of the vector lengths.

$$\cos(\theta) = \frac{\vec{A} \bullet \vec{B}}{\|\vec{A}\| \times \|\vec{B}\|} \tag{8.1}$$

Figure 8.1 plots the cosine distance computed for a document at each sentence. There are 198 sentences in this document, and there are six stories. The boundaries of the stories are marked by vertical dotted line. For each sentence, we compute the cosine distance between two neighboring blocks of sentences: the preceding 20 sentences and the following 20 sentences. It is obvious that the cosine distance achieves local minima near the story boundaries. With a simple threshold method, story boundaries can be detected.

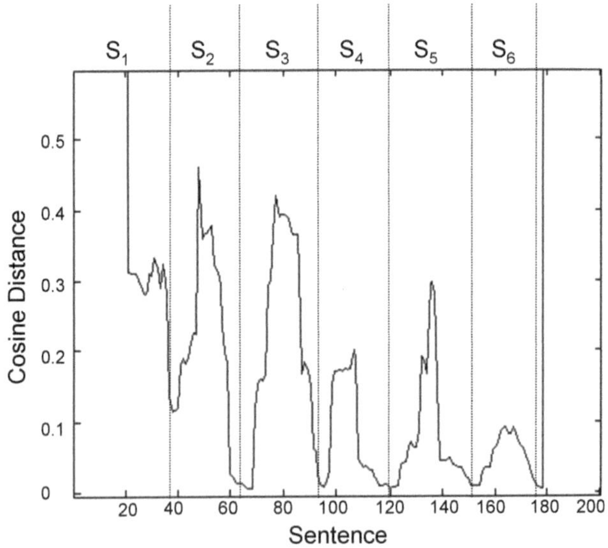

Fig. 8.1. Cosine distance of neighboring text blocks for a news program.

8.2.3 Dynamic Programming

Topic segmentation may be based on multimodal cues, but here we present a method that consists of the following steps and is entirely based on the closed captioned text or automatic speech recognition of the audio component of the video programs:

Input: a set of sentences corresponding to the program dialog transcription for a program unit between commercial breaks, typically from the processed closed caption. For non-commercial content, the entire program text may be used, with slightly lower accuracy.

1. Use a part-of-speech tagger to mark all nouns.
2. Stem all nouns to their roots.
3. Define a symmetric matrix S such the element $S(i, j) = 1$ if sentences i and j have at least one noun in common, otherwise zero. Figure 8.2 shows the S matrix for a half hour news program with 198 sentences, where only pairs of sentences that are less than 100 sentences away are considered. The real story boundaries are marked by dashed lines. There are six stories: S_1 to S_6 in this program, and it is obvious that the S matrix is dense within each story block.
4. Define a density $D(i, j)$ between sentences i and j:

$$D(i,j) = \frac{\sum_{m=i}^{j-1} \sum_{n=m+1}^{j} S(m,n)}{(j-i+1)^r} \tag{8.2}$$

where the exponent r is obtained by cross validation. The numerator is a count of the number of 1's in the upper triangular matrix between sentences i and j and therefore bounded by elements $(i, i+1)$, (i, j) and $(j-1, j)$.

5. Find the set of sentences (i_1, j_1), (i_2, j_2), …, (i_k, j_k), …, (i_K, j_K) with $j_k > i_k$ and $j_k+1 = i_{k+1}$, such that the following is maximized,

$$J = \sum_{k=1}^{K} D(i_k, j_k) \tag{8.3}$$

J can be found by dynamic programming and basically finds the K sets of sentence intervals (i_k, j_k) such that the sum of the densities over these K intervals is maximized.

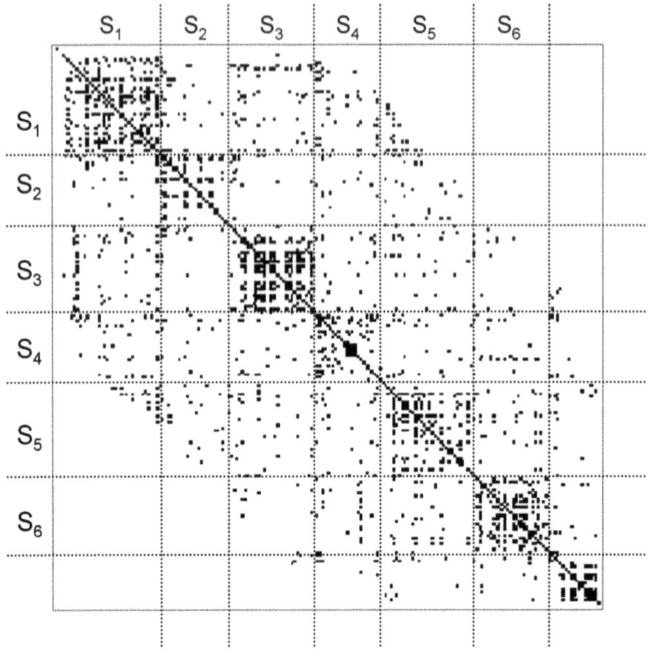

Fig. 8.2. Sentence similarity matrix *S* for a news program.

The algorithm has been developed and tested using the LCD TDT-3 dataset which includes closed caption data from CNN, NBC, ABC, and PRI and includes topic boundary indications.

This process assumes that a good algorithm exists to determine sentence boundaries and this is true for the closed captions in the news programs which typically include end of sentence punctuations. The dynamic programming algorithm is embedded with a constraint so that topic segments are at least three sentences long.

This procedure is similar to [Fragkou04] except they used an additional penalty term if the segment lengths are too long or too short. The attributes "too long" or "too short" are based on experimental analysis of the average length and standard deviations of segment lengths. In addition, they used all words, while this method used nouns only.

The algorithm works better if the program segments are as short as possible. Hence, it is desirable that the segments be the sentences between commercial boundaries. While it is true that sometimes coherent segments of text cross commercial boundaries, the text before the commercial is typically a transitional phrase such as "Coming up next..." and this can be readily detected and removed from the optimization using context-free grammars or other natural language processing techniques.

8.2.4 Topic Classification

BBN Technologies built a topic classification component, called OnTopic, which is based on probabilistic HMM [Makhoul00]. OnTopic is trained to classify thousands of topics. For text based story segmentation, OnTopic with 5500 topic models is applied to overlapping data windows of 200 words with a step size of four words between adjacent windows. For a data window, the top scoring 100 topics are selected to train a Gaussian model to obtain the likelihood values. Then, only those topics whose log likelihood values are more than twice the standard deviation from the mean score are kept as the pruned topic lists for the data window. The next task is to locate the story boundaries based on the pruned topic lists of all data windows. There are two steps involved which first roughly locate the boundaries and then precisely pinpoint the boundaries.

A topic window is defined as the aggregate of 50 consecutive pruned topic lists, and topic persistence is the number of occurrences of each topic label found in a topic window. Then the maximum-persistence scores within each topic window are computed. Within the same story, normally the maximum-persistence value is 50. At the boundaries of the story, the maximum value decreases. A threshold of 90% of the maximum is used to roughly locate the story boundaries. To precisely pinpoint the story boundaries, topic support words are utilized. The topic support words are those words in a topic window that contribute to the score of one of the pruned topics. They are easily separable into two groups whenever they span a story boundary – one group supports the topics identified in the preceding story, and the other group supports the topics in the succeeding story.

8.3 Named Entity Extraction

Named entity extraction (NEE), also known as named entity recognition, means extracting atomic elements with associated categories, such as person names, locations, phone numbers, identification numbers, etc. For example, in sentence "Tom Smith's phone number is 234-456-789," there are two named entities: "Tom Smith" is a person name, and "234-456-789" is a phone number. It is a basic component for natural language processing applications, such as information retrieval, question answering, and document summarization, etc. [Kobayashi03, Kubala98].

The definition of the categories is application dependent. The set of named entities can be partitioned into two sets – application-independent and application-dependent. Examples of application-independent entities include "phone numbers," "dates," "currency," "credit/calling card numbers." These are found in many different applications. Besides application-independent entities, an application may also need some entities specific to the application, for example names of products and services.

Generally speaking there are two approaches for named entity extraction: rule based and data driven approaches. Rule based approaches rely on linguistic rules, while data driven methods rely on statistical pattern recognition methods including hidden Markov models (HMM), maximum entropy (ME), and support vector machines (SVM). In this section, we will show some examples in both categories.

8.3.1 Rule Based NEE

In rule-based approaches, a grammar in Backus Naur Form (BNF) may be created manually for each named entity [Gupta06]. The creation of a new named entity may involve reusing or extending one of the grammars available in a library of application-independent named entities, or it may involve writing a new grammar from scratch. A library of generic grammars is available for such items as phone numbers, and the library may be augmented with application-specific grammars to deal with account number formats, for example.

As an example, a fragment of a "date" grammar is shown in Fig. 8.3. Note that the terminals of the BNF are of the form X:Y where $X \in V' \bigcup \{\varepsilon\}$, $Y \in \{V' \bigcup TAGS\}$. $TAGS = \bigcup_i \{<t_i>, </t_i>\}$ which is the set of start and end symbols representing the entity types, and V' is the vocabulary.

$$\begin{array}{lcl} DATE & \rightarrow & \begin{array}{c} \varepsilon : \langle month \rangle MONTH \ \varepsilon : \langle / month \rangle \\ \varepsilon : \langle day \rangle DAY \ \varepsilon : \langle / day \rangle \end{array} \\ MONTH & \rightarrow & january : january \mid feburary : feburary \mid \cdots \\ DAY & \rightarrow & first : first \mid \mathrm{sec}ond : \mathrm{sec}ond \mid \cdots \end{array}$$

Fig. 8.3. A simple date grammar.

These grammars are typically regular expressions written in a grammar rule notation. They are compiled into finite-state acceptors whose arcs are labeled with the terminals of the grammars. The two components of the arc labels are then interpreted as the input and the output symbols leading to a finite-state transducer representation. The result of compilation of the previous grammar fragment is shown in Fig. 8.4.

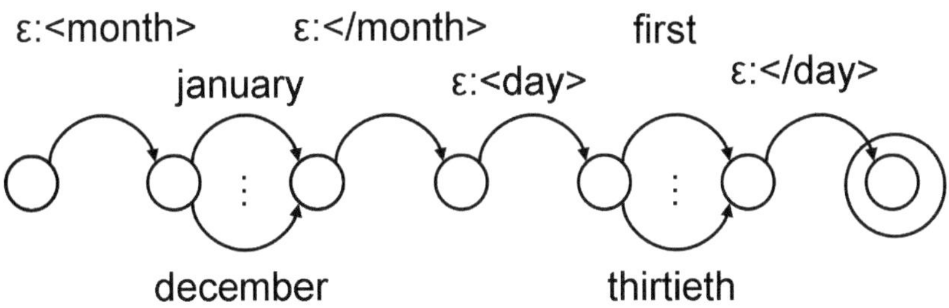

Fig. 8.4. FST representation of the "date" grammar fragment.

Each entity grammar G_i is compiled into a finite state transducer (FST) F_i and the final entity extraction model F is a transducer resulting from a union of all the FSTs: $F = \bigcup_i F_i$. It is often the case that the same substring might represent more than one entity type. An example is a sequence of 10 digits which could be a phone number or an account number. Although, for the majority of named entities of interest, the grammars can specify the context in which the BNF rules can apply, it is clear that this approach is limited and is unable to deal with other ambiguities that cannot be resolved from a small set of immediate contexts.

The kinds of entities we are interested in can be extracted using the procedure discussed previously. Writing accurate grammars with high precision and high recall is a tedious and time consuming activity. In practice, grammars with higher precision are preferred over those with higher recall. If higher recall is needed for an application, data-driven named entity extraction (NEE) can be used.

8.3.2 Data Driven NEE

Data driven NEE relies on two components: feature extraction and the classification method. Typical features include:

- Words and their lemmas in a window surrounding the current word;
- The part-of-speech tags of the current and surrounding words;
- The prefixes and suffixes of the current and the surrounding words;
- *N*-grams;
- Capitalization information.

Gazetteer information, for example, lists of cities, countries, proper names, organizations, etc., and classification results from different approach are also useful features [Kobayashi03].

Maximum entropy methods and support vector machines make decisions for each feature vector, within which the context information has been encoded. The MaxEnt classifier computes the posterior class probability of an example by evaluating the normalized product of the weights active for the particular example. The model weights are trained using the improved iterative scaling algorithm [Borthwick99]. The support vector machine is designed for binary classification tasks, and how to efficiently extend it to multiple classes (for example, the NEE case) is an important algorithm design consideration. Isozaki and Kazawa [Isozaki02] adopted the "one versus all others" approach, where each classifier is trained to distinguish members of a class from no-members. For situations where two or more classifiers each classifies a sample as a member, the distances of the sample to the SVM boundaries are used to determine the best class. Given that the feature dimensionality is normally huge in the text domain, feature selection is useful for further improving the efficiency. Florian et al. [Florian03] explored how to combine multiple data driven classifiers to further improve the NEE performance.

8.3.3 NEE Tools

There are many existing named entity extraction software tools available from public domain. In this section, we give a brief overview of one of them, GATE – General Architecture for Text Engineering [GATE07]. GATE is one of the most widely-used nature language processing systems, and is a very comprehensive infrastructure for language processing software development, implemented in Java. Key features of GATE include (1) Component-based development reduces the system's integration overhead in collaborative research; (2) Automatic performance measurement of language engineering components promotes quantitative comparative evaluation; (3) Distinction between low-level tasks such as data storage, data visualization, discovery and loading of components and the high-level language processing tasks; (4) Clean separation between data structures and algorithms that process human language; (5) Consistent use of standard mechanisms for components to communicate data about language, and use of open standards such as Unicode and XML; (6) Insulation from idiosyncratic data formats (GATE performs automatic format conversion and enables uniform access to linguistic data); and (7) Provision of a baseline set of LE components that can be extended and/or replaced by users as required.

Within the GATE distribution, named entity extraction is included in the information extracting system – ANNIE (A Nearly-New Information Extraction System). ANNIE includes the following components.

1. Tokenizer, which splits the text into very simple tokens such as numbers, punctuation and words of different types.
2. Gazetteer, which contains a set of plain text lists. Each list represents a set of names, such as names of cities, organizations, days of the week, etc.
3. Sentence splitter, which is a cascade of finite-state transducers which segments text into sentences.
4. Part-of-speech tagger, which produces a part-of-speech tag as an annotation on each word or symbol.
5. Semantic tagger, which contains rules that act on annotations assigned in earlier phase, in order to produce outputs of annotated entities.
6. Orthographic coreference, which adds identity relations between named entities found by the semantic tagger, in order to perform coreference.
7. Pronominal coreference, which perfoms anaphora resolution using the JAPE grammar formalism. JAPE is a Java annotation patterns engine.

8.4 Part-of-Speech Tagging

Part-of-speech (POS) tagging means associating words in text to a particular part of speech, such as nouns, verbs, adjectives, adverbs, pronouns, prepositions, conjunctions, and interjections. For example, given the sentence "The kid is smart," the POS tagger would output "The/DT kid/NN is/VB smart/JJ." (See Table 8.1 for definitions of the acronyms.) Tagging text with parts-of-speech is extremely useful for more complicated NLP tasks such as parsing and machine translation.

A big challenge in POS tagging is to solve the tag ambiguities. For example, the word "book" can be a noun in the sentence "I have a book", or a verb in the sentence "I will book a hotel." Most English words are unambiguous, but many of the most commonly used words are ambiguous, which makes POS tagging difficult. In spite of the challenges, state-of-the-art POS taggers can achieve accuracy as high as 96%.

The actual set of tags used in POS taggers is more complex than the general eight types of POS described in the previous paragraph. There are three commonly used training datasets or tagsets: the Brown tagset, the

Penn Treebank tagset, and the C5 tagset. The Brown corpus was created at Brown University in the 1960s, and it collected one million word of samples from 500 written texts of different genres. The Brown tagset used by Brown corpus defined 87 tags. The Penn Treebank tagset is smaller, with 45 tags, while the C5 tagset defines 61 tags. Table 8.1 shows the Penn Treebank tagset.

Table 8.1. Tagset of Penn Treebank.

Tag	Description	Tag	Description
CC	Coordinating conjunction	SYM	Symbol
CD	Cardinal number	TO	To
DT	Determiner	UH	Interjection
EX	Existential there	VB	Verb, base form
FW	Foreign word	VBD	Verb, past form
IN	Preposition or subordinating conjunction	VBG	Verb, gerund or present participle
JJ	Adjective	VBN	Verb, past participle
JJR	Adjective, comparative	VBP	Verb, non-3rd person singular present
JJS	Adjective, superlative	VBZ	Verb, 3rd person singular present
LS	List item marker	WDT	Wh-determiner
MD	Modal	WP	Wh-pronoun
NN	Noun, singular or mass	WP$	Possessive wh-pronoun
NNS	Noun, plural	WRB	Wh-adverb
NNP	Proper noun, singular	$	Dollar sign
NNPS	Proper noun, plural	#	Pound sign
PDT	Predeterminer	“	Left quote
POS	Possessive ending	”	Right quote
PRP	Personal pronoun	(	Left parenthesis
PRP$	Possessive pronoun	)	Right parenthesis
RB	Adverb	,	Comma
RBR	Adverb, comparative	.	Sentence-final punc
RBS	Adverb, superlative	:	Mid-sentence punc
RP	Particle		

Generally speaking, there are two classes of POS tagger: rule-based taggers and stochastic taggers. Rule based taggers normally contain two steps where the first step assigns all possible POS taggers to each word based on a dictionary, and the second step removes the wrong tags based on a large set of disambiguation rules. EngCG [Voutilainen99] is a sample rule-based tagger, which has 3744 constraints and utilizes probabilistic constraints and other syntactic information.

Stochastic taggers compute the probability of a given word in a context for certain tag. We use the hidden Markov model (HMM) based tagger as an example in this category. Given the observation of a string of words $\{w_1, w_2, \ldots, w_n\}$, we need to find the sequence of tags $\{\hat{t}_1, \hat{t}_2, \ldots, \hat{t}_n\}$ that maximize the a posteriori probability $P(\{t_1, t_2, .., t_n \mid w_1, w_2, \ldots, w_n\})$. For simplicity purposes, we denote $\{\hat{t}_1, \hat{t}_2, \ldots, \hat{t}_n\}$ by $\hat{t}_1^n$. Similarly, $\{t_1, t_2, \ldots, t_n\}$ is denoted by t_1^n, and $\{w_1, w_2, \ldots, w_n\}$ by w_1^n. Then,

$$\hat{t}_1^n = \arg\max_{t_1^n} P(t_1^n \mid w_1^n)$$

By Bayes' rule, this can be expressed as

$$\hat{t}_1^n = \arg\max_{t_1^n} \frac{P(w_1^n \mid t_1^n)P(t_1^n)}{P(w_1^n)} = \arg\max_{t_1^n} P(w_1^n \mid t_1^n)P(t_1^n)$$

Here $P(w_1^n \mid t_1^n)$ is the likelihood of string w_1^n given tag sequence t_1^n, and $P(t_1^n)$ is the prior probability of tag sequence t_1^n. The HMM tagger greatly reduces the optimization problem by two assumptions: (1) the likelihood of each word w_i only depends on its tag t_i; and (2) the probability of current tag t_i only depends on its previous tag t_{i-1}. With these two constrains, the equation can be written as

$$P(w_1^n \mid t_1^n)P(t_1^n) = \prod_{i=1}^{n} P(w_i \mid t_i) \cdot P(t_1) \prod_{i=2}^{n} P(t_i \mid t_{i-1})$$

As shown in this equation, the parameters of the HMM are (1) the initial tag probabilities $P(t_1)$, (2) the tag transition probabilities $P(t_i \mid t_{i-1})$, and (3) the word likelihoods $P(w_i \mid t_i)$. These set of parameters can be trained using a large corpus of labeled data.

Inspired by both the rule-based and stochastic based taggers, Brill [Brill93] proposed a transformation-based tagger, which is also called the Brill tagger. The Brill tagger relies on a set of tag rules that are automatically trained from a corpus. It has three stages. In the first stage, every word is labeled with its most likely tag. In the second stage, it checks all possible transformations, and selects the one that leads to the most improvement. In the third stage, data is re-tagged based on this rule.

8.5 Capitalization

Correct text capitalization is an important factor in determining the quality of the transcripts obtained from closed captioned text (which is usually in all upper case) and those generated by automatic speech recognition engines. Following are two sentences; one is the original closed caption extracted from broadcast programs, and the other one is the same sentence with case information restored.

- IT'S "THE TONIGHT SHOW" WITH JAY LENO -- FEATURING KEVIN EUBANKS AND THE "TONIGHT SHOW" BAND. AND I'M JOHN MELENDEZ.
- It's "The Tonight Show" with Jay Leno -- featuring Kevin Eubanks and the "Tonight Show" band. And I'm John Melendez.

The second form is obviously easier to read and demonstrates repurposing of closed caption data for applications such as creating printed transcripts. From a syntactic perspective, capitalization is usually used to indicate the beginning of a new sentence. This is probably the most common usage, but it carries no semantic content. The other uses of capitalization are meant to emphasize the sentient nature of an entity or being the specific work of a sentient being or a specifically named natural location or phenomenon. In fact, it could be considered an insult in some contexts to leave a word non-capitalized. That's because capitalization may also indicate a certain amount of dignity or honor. Because capitalization has such semantic, pragmatic, and sociolinguistic implications, it is helpful to read documents that have correct capitalization.

Closed caption (CC) documents and Telephone Typed Dictation (TTD) are made in a hurry and mostly lack correct capitalization and punctuation. Automatic Speech Recognition (ASR) likewise needs all of its processing directed towards determining the correct word to use for a given sound pattern. Capitalization is an afterthought when dealing with ASR.

Several researchers investigated the text capitalization problem. Chelba and Acero proposed a maximum entropy based capitalizer in [Chelba04], and Brown and Coden proposed an *N*-gram based capitalization method in [Brown01]. To illustrate some of the issues involved in more detail, we will present the case restoration module of MIRACLE system (Multimedia Information Retrieval by Content) [Gibbon06, Liu06] which is a combination of rule-based capitalization and an *N*-gram language model generated using a large corpus of AP newswire data collected back in the 1990s. Keeping this data up-to-date requires timely discovery and mining of recently published documents to learn new information and incorporate it into the models. One promising approach is to crawl through Website content using RSS (Rich Site Summary) feeds, and automatically update the case restoration module based on collected Web data.

8.5.1 Linguistic Processing Architecture

Figure 8.5 shows the system architecture. RSS feeds are lightweight XML files that are regularly maintained by the Websites for sharing new content. The Web document collector periodically queries a set of selected RSS feeds and downloads the news stories following the embedded links. Updated RSS feeds are stored in a Database to track their changes, such that only new content will be fetched. The news stories are stored in a standard database for post processing. The textual information extractor module locates the useful news story segments buried in the complicated HTML pages which normally contain a variety of other information, including advertisements. The plain text news stories are saved in the Web story database.

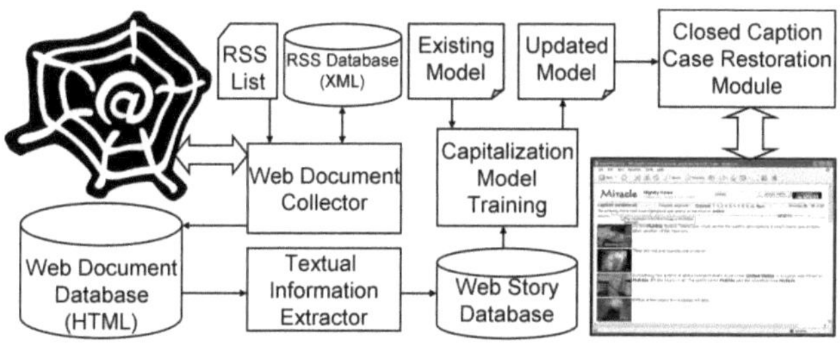

Fig. 8.5. System architecture of the case restoration module.

The capitalization model maintains a set of *N*-grams based on a collection of recent stories, and the new *N*-grams are merged with the existing ones in the case restoration module.

8.5.2 Web Document Collection

An explained in Chap. 2, an RSS file is an XML dialect for Web syndication used by news Websites and Weblogs [RSS07]. Its main purpose is to allow Internet users to subscribe to certain Websites, whose content is regularly updated. Each RSS feed contains a set of items embedded with associated URLs and related metadata.

This system utilizes the RSS feeds of the major broadcast and media publishing companies, for example, ABC, BBC, CBS, Fox, CNN, MSNBC, *The New York Times*, etc. Each company provides dozens of RSS feeds which focus on different news categories, including world news, domestic news, politics, science, health, finance, etc. The wide range of the content at these Websites was chosen to cover the same content domain that the MIRACLE system is processing. Currently, 113 RSS feeds from the above Websites are chosen, and the Web document collector module fetches each of them every two hours to track new items that were recently published. Given that the size of the RSS feeds is reasonably small, all downloaded RSS feeds are saved in the RSS database, and the key for each entry is the RSS feed URL combined with the fetch time.

As we indicated, each RSS feed contains a group of items, which specify the story URLs as well as some metadata. The Web document collector module extracts the URL from each item, and downloads the webpage of the corresponding news story. The Webpages are saved in the Web document database without any processing. Each story is assigned with a key, which is actually the URL of the Webpage, such that the system can skip the items that exist in the database, and just download the new ones. This is especially useful in the case that the same story may appear in more than one RSS feed from the same company, for example, a news story may be listed in both the domestic RSS feed and the politics RSS feed. The metadata provided in the RSS feed for each item as well as the fetching time information are saved in the database also.

On average, more than half of the content in the Webpage is not related to the embedded news story. This includes dynamic HTML controls, programming scripts, HTML tags, advertisements, etc. To extract useful information, the textual information extractor module employs an HTML parser to sift the pure text segments of news stories. To ease the training process, each story is further segmented into a set of sentences using a rule based sentence parser. The parser relies on punctuation information and a set of rules of titles and acronyms, which is primitive yet effective. All sentences are saved in the database. Within the database, data collected from different months are saved separately, such that only the recent data will be used for training the capitalization models.

8.5.3 Text Capitalization Algorithm

The current implementation in the MIRACLE system depends on an *N*-gram language model trained from an older broadcast news corpus. The *N*-gram (up to 9-gram) in this model is simply a sequence of lower case words, also called tokens. These *N*-grams are stored in a hash table, and their values are the target capitalized version of the token. For example, Hash("big apple") = "Big Apple". When the capitalizer runs, it looks for the longest *N*-gram in the current position in the text that matches, and then capitalizes it as it's capitalized in the *N*-gram. The algorithm starts from $N = 9$ words from the current position, and if it doesn't match it will fall back to a strategy of looking at only $N-1$ words from the current position, all the way until only the single word at the current position in the text is examined to see if it exists capitalized in the *N*-gram hash. If it doesn't, then the word isn't capitalized. In any case, the algorithm will move 1 to N words ahead once it capitalizes as much as it can from the current position.

The capitalization model training module retrieves all news story sentences of the most recent three months. For training purposes, the first word of each sentence is ignored, which is always capitalized. The training process contains two major steps. In the first step, each chunk of adjacent capitalized terms is identified, and the lower case of the chunk is used as a token. For example, token_1 = "big apple." The length of a token varies from a single term to nine terms. At the second step, the entire corpus is scanned again to compute the statistics of all possible capitalizations of each token. For example, the statistics of token_1 in one training corpus is following,

- Frequency("big apple") = 37,
- Frequency("big apple" => "big apple") = 2,
- Frequency("big apple" => "Big Apple") = 35.

To get rid of possible typos, the system ignores those tokens whose total occurrence is less than three times. If a certain type of capitalization is significant enough, which means that its frequency is dominant compared to all other alternatives (e.g. more than 70% of the token frequency), it is recorded as one *N*-gram. The *N*-gram created in the above example is Hash("big apple") = "Big Apple".

Now a list of *N*-grams is built based on the training corpus, and the redundant *N*-grams need to be removed. For example, if the *N*-gram list contains the following three tokens:

- Hash("tom smith") = "Tom Smith"
- Hash("tom") = "Tom"
- Hash("smith") = "Smith"

It is clear that the first *N*-gram is redundant with respect to the last two *N*-grams, and in this case, the first *N*-gram is removed to reduce the processing complexity. After the set of *N*-grams is built from the new training corpus, they are merged with the existing *N*-grams. The merging process is composed of two steps: First, for each token in the new *N*-grams, the system either updates the existing *N*-gram if the token exists or adds a new *N*-gram if the token is new. Second, the redundancy is removed in the merged *N*-grams. The merged *N*-grams are used by the case restoration module for any new content acquired in MIRACLE system.

8.6 Information Retrieval

8.6.1 Stemming

Stemming is the process of reducing derived words to their stem form, which need not be identical to the morphological root of the word. It is usually sufficient that related words map to the same stem. In English, a verb has a number of morphological forms. For example, the non-third-person-singular (eat), third-person-singular (eats), progressive (eating), past participle (eaten). Stemming maps eat, eats, eating, eaten into the same stem, eat.

The Porter stemmer is a widely used method, which automatically removes the suffix based on a set of rules. Details of the method can be found in [Porter80]. Here we briefly introduce the method. The algorithm first defines consonant (c) as letters other than A, E, I, O, U, or Y preceded by a consonant. If a letter is not a consonant, it is a vowel (v). A list of consonants of length greater than 0 is denoted by C, and a list of one or more vowels is denoted by V. Then, any words can be represented by [C](VC){m}[V], where the square brackets denote arbitrary presence of their contents, and (VC){m} means VC repeats m times.

The Porter stemmer defines a set of rules to remove a suffix. These rules are in the form of "(condition) S1 → S2," which means if a word ends with suffix S1, and the stem before S1 satisfies the condition, S1 is replaced by S2. S2 can be null. The algorithm applies five steps of rules sequentially to strip complex suffixes for a given word. The first step deals with plurals and past participles. A few rules and corresponding examples are listed below:

SSES → SS	caresses → caress
(m>0) EED → EE	agreed → agree

There are about a dozen rules in the first step, and only one with longest matching S1 is obeyed. The next four steps are more straightforward, each containing a set of rules. We give one example for each of these steps,

Step 2:	(m>0) ATIONAL → ATE	relational → relate
Step 3:	(m>0) ICATE → IC	triplicate → triplic

Step 4:	(m>1) AL → null	revival → reviv
Step 5:	(m=1 and not *o) E → null	cease → ceas

Going through these sets of rules for each input word is time consuming. A more efficient implementation is to use a dictionary (e.g. a Hash table built on an existing corpus by these rules) to map a known word into its stem. Only unknown words have to go through the five steps of rules.

8.6.2 Term Weighting

Term frequency is the number of times that each term appears in a document and it is a useful feature for text processing. Obviously, simply using term frequency to represent a document is not effective, since generally, some words appear more often than the other words. For some most common words, e.g. the, a, at, etc., it is good idea to remove them before further text processing. These words are called stop words or noise words. Term weighing is necessary to emphasize that more information is brought in by less common words.

TF-IDF (Term Frequency – Inverse Document Frequency) [Salton88] weighting is an ad hoc modification to the cosine coefficient calculation which weights words according to their usefulness in discriminating documents. Words that appear in few documents are more useful than words that appear in many documents. This is captured in the equation for the inverse document frequency of a word:

$$idf(w) = \log\left(\frac{N}{df(w)}\right) \tag{8.4}$$

where $df(w)$ is the number of documents in a collection which contain word w, and N is the total number of documents in the collection. The resulting similarity measure between two documents a and b is:

$$sim(a,b) = \frac{\sum_{w=1}^{n} tf_a(w) \times tf_b(w) \times idf(w)}{\sqrt{\sum_{w=1}^{n} tf_a^2(w)} \times \sqrt{\sum_{w=1}^{n} tf_b^2(w)}} \tag{8.5}$$

8.6.3 Ranking

After the retrieval engine finds all relevant results for a query, it needs to rank them in a certain order before presenting them to the user. One common ranking method is based on the creation time of the document. In such cases, the user can choose to see the newest document first or browse the oldest document first.

In certain cases, ranking based on time is not the most desirable method for a user. While a user performs a search, he/she wants to see the most relevant results first. For example, the relevance here can be measured by the number of occurrences of the queried terms or the density of the queried terms in the retrieved documents. Here, the density of a term is defined as the ratio of the times that the term occurred to the length (in words) of the document. Obviously, the TF-IDF introduced in the previous section is an effective measurement of relevance as well.

For Web document retrieval, a well known ranking method is called PageRank invented by Page [Page06], who co-founded Google. Assuming there are N documents $p_1, \ldots, p_N$, the page rank of document p_i, denoted by $PR(p_i)$, is determined by the following equation,

$$PR(p_i) = \frac{1-d}{N} + d \sum_{p_j \in M(p_i)} \frac{PR(p_j)}{L(p_j)}, \tag{8.6}$$

where $M(p_i)$ is the set of documents that link to p_i, $L(p_j)$ is the number of outbound links in document p_j, and d is a damping factor, which can be set around 0.85. Let us use vector $\mathbf{R}$ to represent the PageRank vector $[PR(p_1)\ PR(p_2) \ldots PR(p_N)]'$, the previous formula can be written in matrix format,

$$\mathbf{R} = \begin{bmatrix} (1-d)/N \\ (1-d)/N \\ \vdots \\ (1-d)/N \end{bmatrix} + d \begin{bmatrix} l(p_1,p_1) & l(p_1,p_2) & \cdots & l(p_1,p_N) \\ l(p_2,p_1) & l(p_2,p_2) & \cdots & l(p_2,p_N) \\ \vdots & \vdots & \ddots & \vdots \\ l(p_N,p_1) & l(p_N,p_2) & \cdots & l(p_N,p_N) \end{bmatrix} \mathbf{R} \tag{8.7}$$

where $l(p_i, p_j)$ is 0 if document p_j has no link to p_i, otherwise, a normalized constant such that $\sum_{i=1}^{N} l(p_i, p_j) = 1$.

An iterative procedure can be used to determine (or approximate) the value of $\mathbf{R}$. The PageRank of each document $PR(p_i)$ is set to $1/N$ at the beginning, and good results are achieved after a few iterations.

8.7 Text Summarization

With the rapid growth of the World Wide Web and computer based information systems, information is becoming available on-line at an increasing rate. One consequence is the so called information explosion. No one has time to read everything, yet we often have to make critical decisions based on what we are able to assimilate. The technology of automatic text summarization is a solution to this problem. Text summarization is the process of distilling the most important information from a source to produce an abridged version for a particular user or task. Text summarization has been a research topic since the 1950s, however, it became more active since the second half of the 1990s due to the boom of the Internet.

A summary is a concise restatement of the topic and main ideas of its source. Generally speaking, there are three stages in automated text summarization: (1) topic identification, (2) interpretation or topic fusion, and (3) summary generation [Mitkov05]. Topic identification is to determine the central topic(s) for the input text. Normally, this is through first parsing the input text into syntactic and semantic representations, and then analyzing the relations between these representations and various topics to choose the most relevant topic(s) discussed in the text. Topic interpretation and fusion refers to the process of fusion of identified topics and expressing the subject using concepts or words that may not occur in the original text. This stage requires domain knowledge that may not exist in the input text. Summary generation is to create the summary content through natural language generation techniques based on the abstract and information extracted in the previous two stages. The length of the final summary varies from a few paragraphs to a couple of sentences. The extreme cases include a list of keywords extracted from the text.

The TIPSTER Text Program was a Defense Advanced Research Projects Agency (DARPA) led government effort to advance the state of the art in text processing technologies through the cooperation of researchers and developers in government, industry and academia [Tipster07]. In its efforts to improve document processing efficiency and cost effectiveness TIPSTER focused on three underlying technologies.

- Document Detection: the capability to locate documents containing the type of information the user wants from either a text stream or a store of documents.
- Information Extraction: the capability to locate specified information within a text.
- Summarization: the capability to condense the size of a document or collection while retaining the key ideas in the material.

The text summarization task in the TIPSTER Text Program provides a common platform for all participating research groups to develop novel text summarization technologies. Some of the techniques that may be used, independently or combined, in building summaries include [Tipster07]:

- Selecting important paragraphs from a document.
- Selecting important sentences from a document.
- Selecting high frequency, meaningful words from a document.
- Selecting unusual words from a document.
- Counting repeated word usage to identify important sentences.
- Using information extraction techniques to identify important document entities, e.g. person names, place names, company names, organizations, numeric data and temporal data.
- Using vector techniques to group either documents or paragraphs under common concepts.
- Using retrieval techniques to identify documents that correspond to a complex query which would be the desired summary.
- Performing some level of modification of the selected sentences or paragraphs using natural language or statistical techniques.
- Using natural language techniques to synthesize new sentences or paragraphs.
- Applying statistical techniques to condense documents or collections of content.

It should be noted that effective summarization is not an easy task and it frequently involves semantic analysis and applying world knowledge for the clearest presentation. While some research has been done and a few trial systems have been developed there is still much, much work to be performed before really good summarization systems become available.

The goal of text summarization can be said to obtain a good summary, but it has been thought difficult to evaluate summaries which are the outputs of text summarization systems, and we do not have definite standard measures to evaluate such systems. Here we introduce two commonly used measures: compression ratio (CR) and information retention ratio (RR) [Mitkov05]. Assume the original text document is T, and its summarization is S, then,

$$CR = \frac{length\ of\ S}{length\ of\ T}; \quad RR = \frac{info\ in\ S}{info\ in\ T} \tag{8.8}$$

A perfect summarization system is supposed to keep all information – a unitary value of RR, with a minimum length of words – a value of CR close to zero. CR is straightforward to compute, but the evaluation of RR may subject to experts' judgment.

While summarizing a single text is difficult enough, summarizing a collection of thematically related documents poses several additional challenges. In order to avoid repetition, one has to identify and locate thematic overlaps. One also has to decide what to include for the remainder, to deal with potential inconsistencies between documents, and, when necessary, to arrange events from various sources along a single timeline. For these reasons, multi document summarization is much less developed than its single-document counterpart, and multi lingual considerations just further increase the difficulty.

In this section, we present a simple text summarization method, where a set of key words or phrases are extracted as a summary for a document.

8.7.1 Keyword Extraction

Often a list of representative key phrases (including keywords) which serve as a dense summary for a document can effectively convey the essence of the document to the user. Keywords have been widely used for indexing and retrieval of documents in databases, especially large ones. In the case of presentation slides, they can also help to rank a slide's relevance to a query. Johnston et al. [Johnston07] extract a list of key phrases with importance scores for each document, and key phrases from a set of documents can be merged and ranked based on their scores.

There are different ways to automatically extract keywords for a text document within a corpus. A popular approach is to select keywords that frequently occur in one document, but do not frequently occur in the rest of the documents based on the term frequency–inverse document frequency (TF-IDF) feature. The difference here is in choosing key phrases for a single document, independent of the other documents. Accordingly, Johnston et al. adopted a different feature, which is term frequency–inverse term probability (TF-ITP). The term probability measures the probability that a term may appear in a general document, and it is a language dependent characteristic.

Assuming that a term T_k occurs tf_k times in a document, and its term probability is tp_k, the TF-ITP of T_k is defined as, $w_{Tk} = tf_k / tp_k$. This method can be extended to assign an importance score to each phrase. For a phrase $F_k = \{T_1\ T_2\ T_3\ \ldots\ T_N\}$, which contains a sequence of N terms, assuming it appears ff_k times in a document, its importance score, IS_k, is defined as,

$$IS_k = \sum_{i=1}^{N} \frac{ff_k}{tp_i} \tag{8.9}$$

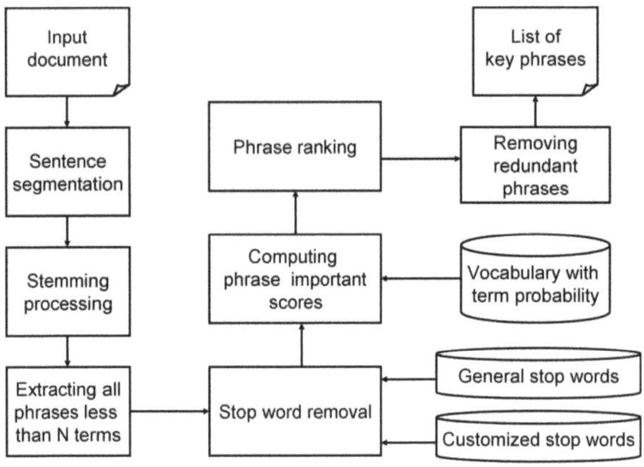

Fig. 8.6. Illustration of key phrase extraction.

Figure 8.6 illustrates the key phrase extraction approach in [Johnston07]. The first step is to segment the input document into sentences based on the punctuations and a set of heuristic rules. For each sentence, the system first applies the Porter stemming algorithm [Porter80] to get rid of word variations, and then extracts all phrases up to N terms long, where N is equal to 4. All phrases that either start or end with noise words are removed. Next the system assigns an importance score for each phrase using estimated term probabilities from a vocabulary based on transcripts of 600 hours of broadcast news data. If a term in the document is out of the vocabulary, and its term frequency is more than 2, then a default term probability value tp_d will be used. The value of tp_d is the minimum term probability in the vocabulary. Finally, phrases are sorted based on their scores, such that the phrases with high scores are chosen as key phrases. Within this step, all phrases that are part of any phrases with higher scores are removed.

An overall list of key phrases for a set of documents is created by merging the individual key phrase lists and summing the importance scores of repeated key phrases. Again, the phrases that are part of any phrase with higher overall score are removed. In this system, the top 10 phrases are kept in the final overall key phrase list.

8.8 Conclusion

In this chapter, we introduced some fundamentals in text processing that are relevant to content analysis, information extraction, and information retrieval. Specifically, we introduced part of speech tagging, named entity extraction, text capitalization, stemming, term weighting, and document ranking. We also presented a few methods for story segmentation and text summarization.

References

[Allen95] Allen, J.: *Natural Language Understanding*. Benjamin/Cummings (1995).

[Baeza99] Baeza-Yates, R. and Ribeiro-Neto, B.: *Modern Information Retrieval*. ACM Press (1999).

[Borthwick99] Borthwick, A.: A maximum entropy approach to named entity recognition. PhD thesis, New York University (1999).

[Brill93] Brill, E.: A corpus-based approach to language learning. PhD thesis, University of Pennsylvania (1993).

[Brown01] Brown, E.W. and Coden, A.R.: Capitalization recovery for text. *Proc. of the SIGIR* (2001).

[Chelba04] Chelba, C. and Acero, A.: Adaptation of maximum entropy capitalizer: Little data can help a lot. *Proceedings of the Conference on Empirical Methods in Natural Language Processing* (2004).

[Feng03] Feng, d., Siu, W., Zhang, H.: *Multimedia information retrieval and management: Technological fundamentals and applications.* Springer, (2003).

[Florian03] Florian, R., Ittycheriah, A., Jing, H., Zhang, T.: Named entity recognition through classifier combination. *Proc. of CoNLL*, pp. 56–62 (2003).

[Fragkou04] Fragkou, P., Petridis, V., Kehagias, A.: A dynamic programming algorithm for linear text segmentation. *Journal of Intelligent Information Systems*, **23**(2), pp. 179–197 (2004).

[GATE07] GATE, A General Architecture for Text Engineering, http://gate.ac.uk/, cited 10 Dec 2007.

[Gibbon06] Gibbon, D., Liu, Z., Shahrary, B.: The MIRACLE video search engine. *CCNC* (2006).

[Gupta06] Gupta, N., Tur, G., Hakkani-Tür, D., Bangalore, S., Riccardi, G., Gilbert, M.: The AT&T spoken language understanding system. *IEEE Transactions On Audio, Speech, and Language Processing*, **14**(1), pp. 213–222 (2006).

[Isozaki02] Isozaki, H. and Kazawa, H.: Efficient support vector classifiers for named entity recognition. *Proc. of COLING*, pp. 390–396 (2002).

[Johnston07] Johnston, M., Ehlen, P., Gibbon, D., Liu, Z.: The multimodal presentation dashboard. Proceedings of the Workshop on Bridging the Gap: Academic and Industrial Research in Dialog Technologies (2007).

[Jurafsky00] Jurafsky, D. and Martin, J.H.: *Speech and Language Processing: An Introduction to Natural Language Processing, Computational Linguistics and Speech Recognition*. Prentice Hall (2000).

[Kobayashi03] Kobayashi, A., Och, F.J., Ney, H.: Named entity extraction from Japanese broadcast news. Proc. *of Interspeech–Eurospeech*, pp. 1125–1128 (2003).

[Kubala98] Kubala, F., Schwartz, R., Stone, R., Weischedel, R.: Named entity extraction from speech. Proc. of DARPA Broadcast News Transcription and Understanding Workshop (1998).

[Liu06] Liu, Z., Gibbon, D., Shahrary, B.: Multimedia content acquisition and processing in the MIRACLE system. *CCNC* (2006).

[Makhoul00] Makhoul, J., Kubala, F., Leek, T., Liu, D., Nguyen, L., Schwartz, R., Srivastava A.: Speech and language technologies for audio indexing and retrieval. *Proceedings of the IEEE*, **88**(8), pp. 1338–1353 (2000).

[Manning00] Manning, C.D. and Schutze, H.: *Foundations of Statistical Natural Language Processing*. The MIT Press (2000).

[Merlino97] Merlino, A., Morey, D., Maybury, M.: Broadcast news navigation using story segmentation. *ACM Multimedia* (1997).

[Mitkov05] Mitkov, R. (ed): *The Oxford Handbook of Computational Linguistics*. Oxford University Press, (2005).

[Page06] Page, L.: Method for Node Ranking in a Linked Database. United States Patent, US 7,058,628 (2006).

[Porter80] Porter, M.F., An algorithm for suffix stripping. *Program*, **14**(3), pp. 130–137 (1980).

[RSS07] RSS 2.0 Specification, http://blogs.law.harvard.edu/tech/rss, cited 10 Dec 2007.

[Salton88] Salton, G. and Buckley C.: Term-weighting approaches in automatic text retrieval. *Information Processing and Management*, **24**(5), pp. 513–523 (1988).

[Tipster07] NIST's TIPSTER Text Program, http://www.itl.nist.gov/iaui/894.02/related_projects /tipster/condense.htm, cited 10 Dec 2007.

[TREC07] Text Retrieval Conference (TREC), http://trec.nist.gov/, cited 10 Dec 2007.

[Voutilainen99] Voutilainen, A. Hand-crafted rules. In van Halteren, H. (Ed.), *Syntactic Wordclass Tagging*. pp. 217–246 (1999).

[Wayne00] Wayne, C.L., Multilingual topic detection and tracking: successful research enabled by corpora and evaluation, *LREC* (2000).

9 Multimodal Processing

9.1 Introduction

With a multimedia document, its semantics are embedded in multiple forms that are usually complimentary each other. For example, a live report on TV about a tsunami conveys information that is far beyond what we read from the newspaper. Therefore, it is necessary to analyze all types of data: image frames, sound tracks, text that can be extracted from image frames, and spoken words that can be deciphered from the audio track [Wang00]. For some applications, automated techniques that process single media, for example, audio or images, may be error-prone, and multimodal processing is used to improve the overall system accuracy.

Multimedia content processing covers a wide area of research activities. Multimodal speech recognition utilizes lip motion, ultrasound images, and acoustic features to improve the speech recognition accuracy [Chen98]. Boreczky [Boreczky98] used HMM framework for video segmentation using both audio and image features. Saraceno and Leonardi [Saraceno98] considered segmenting a video into the following basic scene types: dialogs, stories, actions, and generic. This is accomplished by first dividing a video into audio and visual shots independently, and then grouping video shots so that audio and visual characteristics within each group follow some predefined patterns. In [Huang98], a hierarchical segmentation approach was proposed that can detect scene breaks and shot breaks. The algorithm is based on the observation that a scene change is usually associated with simultaneous changes of color, motion, and audio characteristics, whereas a shot break is only accompanied with visual changes. Lienhart et al. [Lienhart99] proposed using different criteria to segment a video into scenes with similar audio characteristics such as scenes with similar settings and dialogs. The scheme considers audio features, color features, orientation features, and face information. Fisher et al. [Fisher00] used a nonparametric approach to learn the joint distribution of the visual and auditory signals. This work extends the notion of multimedia fusion to complex domains where the statistical relationship between audio and video is complex and non-Gaussian. Cutler and Davis [Cutler00] exploited the correlation between audio and video to search for speaking persons, where the correlation is learned by a time delayed neural network. Adams et al. [Adams03] described a method for automatic labeling high-level semantic concepts in documentary style videos using audio, text, and visual cues. Information from different modalities is combined using support vector machines (SVM). Name-It [Satoh99] is a project aimed at automatically associating faces detected from video frames and names extracted from closed captions for news video. Besides the difficulties in detecting faces and names, the association of them also poses a challenge since multiple faces may appear in one frame and multiple names may be mentioned in one closed caption sentence.

Li et al [LiD03] used cross-modal association to detect talking heads. Cross-modality information analysis extracts multimedia content by identifying and measuring the intrinsic associations between different modalities. In cross-model information retrieval, queries from one modality are used to search for the content in another modality using low-level features. Li and Kuo [Li03] studied how to employ multiple media cues, including audio, visual and face information to analyze video content. The authors also presented a video abstraction system based on video semantics and video production rules. Snoek and Worring [Snoek05] reviewed the state-of-the-art of multimodal video indexing. They concluded that multimodal analysis is the future, and more attention needs to be given to the following factors: (1) content segmentation; (2) modality usage; (3) multimodal integration; and (4) technique taxonomy.

Existing multimedia management systems employ various multimedia content processing techniques. IBM Research developed a prototype multimedia analysis and retrieval system, called MARVEL [Marvel07]. It consists of two components: a multimedia analysis engine, which applies machine learning techniques to model semantic concepts in video, and a multimedia search engine, which integrates semantics-based searching with other search techniques (speech, text, metadata, audio-visual features, etc.). The Informedia II [Christel05] digital video library at CMU is another pioneering multimedia database system. Informedia combines speech recognition, image understanding and natural language processing technologies to automatically transcribe, segment, index, and summarize the linear video. The current library consists of 1500 hours of video. MIRACLE [Gibbon06, Liu06] is an ongoing research project at AT&T Labs aimed at creating automated content-based media processing algorithms and systems to collect, organize, index, mine, and repurpose video and multimedia information. This video search engine combines existing metadata with content-based information that is automatically extracted from the audio and video components.

This chapter guides the reader through three multimedia processing modules: caption/transcript alignment, multimodal story segmentation, and major cast detection in video. The reader can easily appreciate the necessity and the superiority of multimedia content processing for real world applications.

9.2 Case Studies

9.2.1 Closed Caption Alignment

Closed captioning provides useful information for hearing impaired customers and foreigners who watch TV. Since most of the closed caption is generated by a stenographer on the fly with the airing of TV programs, it is delayed from the actual utterance. The delay can be as long as 10 seconds. In video indexing and browsing systems, video should be played back with the synchronized closed caption. Otherwise, the mismatch between the heard audio and the shown closed caption may be very annoying. Aligning the closed caption with speech is very useful for other applications such as topic segmentation.

In this section, we describe an algorithm for closed caption alignment that is adopted in the MIRACLE system [Gibbon06] at AT&T. Figure 9.1 illustrates the block diagram of the system. Instead of applying the speech recognizer on the entire audio and aligning the recognition results with the closed caption globally, the closed caption is segmented into short pieces and each of them is aligned independently first. This method basically breaks a large, time and memory consuming task into many smaller ones. The input closed caption stream is chopped into sentences by a sentence segmentation tool that mainly relies on punctuations and a set of heuristic rules that cope with acronyms, titles, etc. Each sentence has three properties: starting time, ending time, and text. Then the system utilizes the AT&T Watson Automatic Speech Recognition (ASR) tools to align each sentence with the corresponding audio utterance. A grammar based on sentence text, and a clip of audio based on the extended sentence starting and ending times are fed in the ASR module. After compiling the grammar, the Watson ASR utility determines the actual starting and ending times of the corresponding sentence using forced alignment.

In perfect situations, the list of new timestamps for all sentences is accurate and valid. But normally the naïve alignment is not sufficient due to the following two reasons. First, the speech is not clean, and ASR fails to recognize the corrupted utterances. Second, the caption text does not match the real words in speech. This may happen very often due to the typos in closed caption, the causal words spoken by the news hosts that are neglected in closed caption, and for on-screen text. All these possibilities hurt the robustness of the plain alignment algorithm, and it takes extra effort to cope with the detrimental effects in the post processing.

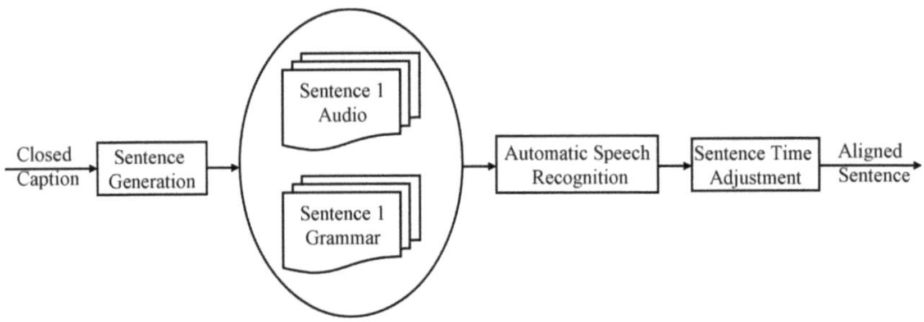

Fig. 9.1. Diagram of closed caption alignment method.

Basically the problem that needs to be solved is to identify the conflicts among aligned timestamps of all sentences and reasonably adjust them. For example, the aligned starting time of one sentence is earlier than the aligned ending time of previous sentence. When no conflicts exist, it does not mean the alignment results are accurate, but there is not much to correct either. When conflicts happen, these indicate mistakes in alignment procedure. A reasonable assumption is that longer sentences have lower possibility of alignment errors, and the timestamps of long sentences can be used to fix the conflicts introduced by neighboring shorter sentences. The following section gives more details of the alignment method.

Alignment algorithm

Suppose there are N sentences $\{S_1, \ldots, S_N\}$, and the original starting and ending times of speaker S_i are τ^i_{st}, and τ^i_{end}. Since the majority of delays lie in the range of 0-8 seconds, it is necessary to extract the audio clip that spans $[\tau^i_{st} - 8, \tau^i_{end}]$ as input for the ASR engine for the current sentence. Let us denote the ASR aligned time for speaker S_i by T^i_{st}, and T^i_{end}. The top part of Fig. 9.2 shows a possible naïve alignment results. Each sentence is marked by two vertical bars, linked by a horizontal bar. The bold solid bar identifies the starting time, and the dashed bar marks the ending time. Due to various reasons, the list of aligned timestamps may be invalid. An example is shown in Fig. 9.2, where the first three aligned sentences overlap. In this section, the focus is on how to rectify such errors and produce valid alignment results.

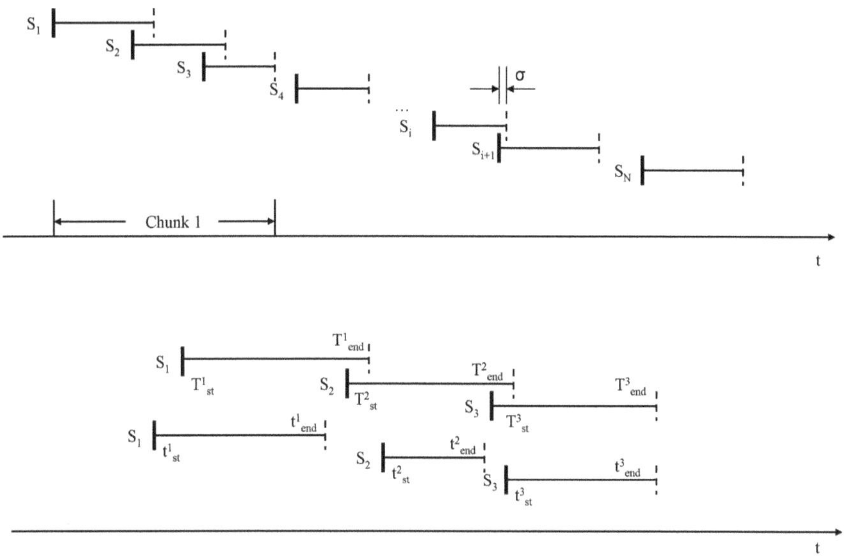

Fig. 9.2. Illustration of timestamp adjustment.

There are two steps in time adjustment. First, if the overlap between adjacent sentences is small, say, less than 0.5 second, then the boundary of the shorter sentence is adjusted to the boundary of the longer sentence. Second, the adjacent conflicting sentences are grouped into isolated chunks, and the timestamps are adjusted within each chunk. For example in Fig. 9.2, the overlap between S_i and S_{i+1} is small, so we adjust the ending time of S_i to the beginning time of S_{i+1} since S_{i+1} is longer. On the other hand the sentence S_1, S_2, and S_3 have longer overlaps among themselves, but do not conflict with others, so they are grouped into a chunk, and adjustment of the timestamps is done within the chunk.

The method of timestamp adjustment within a chunk is shown in the bottom part of Fig. 9.2. One simple solution is to modify the original closed caption time by the average delays. The average delay Δ is computed by the following formula.

$$\Delta = \frac{\sum_{i=1}^{N}\left[\left(\tau_{st}^{i} - T_{st}^{i}\right) + \left(\tau_{end}^{i} - T_{end}^{i}\right)\right]}{N} \tag{9.1}$$

Let $t^i_{st} = \tau^i_{st} - \Delta$, and $t^i_{end} = \tau^i_{end} - \Delta$, we get a valid list of sentences, since the original times from closed caption are valid (no overlap). For each chunk with conflicts, the algorithm tries to replace as many average delayed timestamps t^i_{st} and t^i_{end} by aligned timestamps T^i_{st} and T^i_{end} as possible, so long as the results within the chunk are still valid. This procedure starts from the longest sentence within the chunk, and ends at the shortest one. For the example shown in Fig. 9.2, the original timestamps are $\{(t^1_{st}, t^1_{end}), (t^2_{st}, t^2_{end}), (t^3_{st}, t^3_{end})\}$. For sentence S_1, the system tests whether T^1_{st} and T^1_{end} conflict with the other sentences. Since it does not, the aligned timestamps are used for S_1. The new timestamps of the chunk become $\{(T^1_{st}, T^1_{end}), (t^2_{st}, t^2_{end}), (t^3_{st}, t^3_{end})\}$. This procedure is then applied for S_3 and S_2, and the final timestamps are $\{(T^1_{st}, T^1_{end}), (t^2_{st}, t^2_{end}), (T^3_{st}, T^3_{end})\}$. After all chunks with conflict are rectified, a valid alignment result is achieved.

Simulation Results

The performance of the closed caption alignment algorithm was tested on two NBC Nightly News programs on Jan. 18 and 19, 2000, each half an hour long. Let us denote these two testing data sets as test1 and test2. The first program has 221 sentences, and the second has 203. The real starting time and ending time for each sentence are manually labeled as ground truth.

The histograms of the unaligned starting time offset and ending time offset of all sentences in test1 and test2 are shown in Fig. 9.3(a) and Fig. 9.4(b) respectively. The average boundary difference is 2.7 and 2.9 seconds. The alignment results of the two sequences are shown in Figs. 9.4(a) and (b). The average boundary difference is reduced to 0.2 second and 0.3 second.

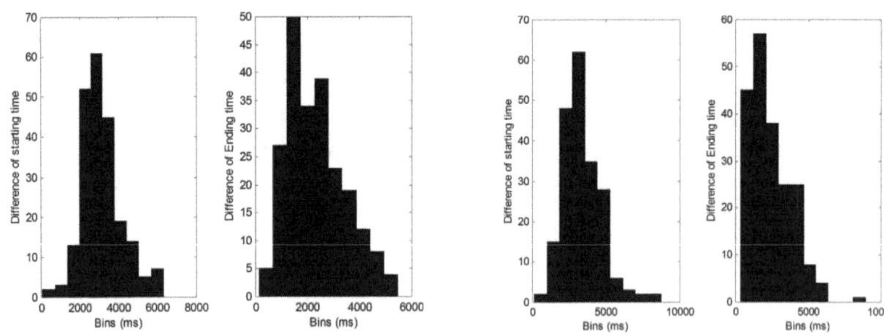

(a) (b)

Fig. 9.3. Histograms of starting time offset and ending time offset between unaligned sentences and ground truth for sequences (a) test1 and (b) test2.

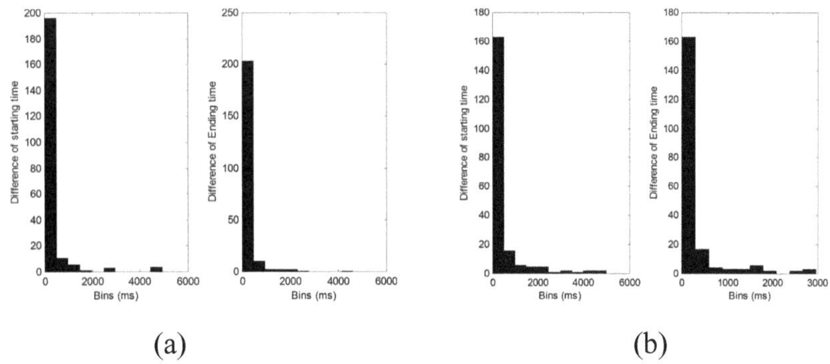

Fig. 9.4. Histograms of starting time offset and ending time offset between aligned sentences and ground truth for sequences (a) test1 and (b) test2.

9.2.2 Multimodal News Story Segmentation

Multimodal news story segmentation algorithms intend to provide users with the ability to retrieve broadcast news programs in a semantically meaningful way at different levels of abstraction. Segmentation algorithms are developed, aimed at automatically generating a content hierarchy as illustrated in Fig. 9.5. The lowest level contains the original multimedia data (audio, video, and text). The next level separates news from commercials. Then the news is segmented into the anchorperson's speech and the speech from others (reporters, interviewees, etc.). Based on this information, higher levels of semantics can be invoked to further segment the data into news stories and news summaries. In turn, each news story can be segmented into an introduction by the anchorperson followed by detailed reporting [Huang99]. With story boundaries detected, video search engine is able to retrieve relevant and complete video segments for users, and additional value added services, such as personalized video query or video alert services can be easily built.

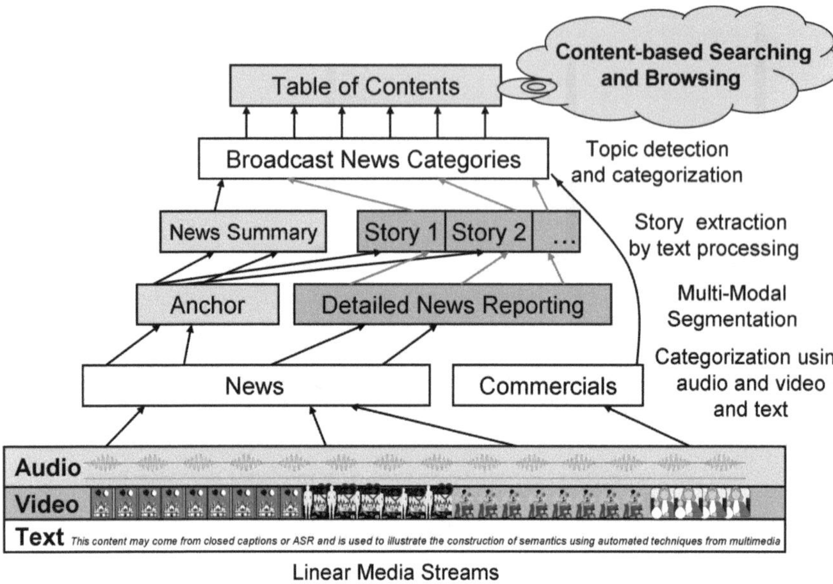

Fig. 9.5. Content hierarchy of broadcast news programs.

A typical news program consists of both news and commercials. News is composed of several headline stories, each of which is usually introduced and summarized by the anchor prior to and following the detailed reporting conducted by correspondents and others. Stories do not typically span commercial boundaries. With this structure of the data, Huang et al. proposed an integrated solution to achieve automatic segmentation of news data into the content hierarchy shown in Fig. 9.5 by utilizing cues from different media.

To separate news from commercials, audio and video information is combined. Within each news segment, the anchorperson's speech is further identified based on speaker detection techniques. Each segment of the anchor's speech is a hypothesized starting point for a new story. The audio-based processing results are then integrated with text-based information processing to obtain higher levels of semantically meaningful abstraction such as stories, story summaries, summary of the day, etc.

News/Commercials Separation Using Audio

News and commercials may be separated based on audio measurements. For example, nine acoustic features are extracted from audio clips: Non-Silence Ratio (NSR), Standard Deviation of Zero crossing rate (ZSTD), Volume Standard Deviation (VSTD), Volume Dynamic Range (VDR),

Volume Undulation (VU), 4 Hz Modulation Energy (4ME), Smooth Pitch Ratio (SPR), Non-Pitch Ratio (NPR), and Energy Ratio in Subband (ERSB). These features are chosen so that the underlying audio events (news vs. commercials) can be reasonably separated in the acoustic feature space. Clip level features are computed from frame level features, where each frame consists of 512 samples and adjacent frames are overlapped by 256 samples and each clip is composed of a set of frames [Liu98]. Three different classification methods were tested in separating news from commercials: linear classifier, fuzzy classifier, and GMM model based classification. Even though the classification is performed on each clip, the precise boundary between news and commercials (which can be in the middle of a clip) is determined by also considering the video processing results: the boundary cannot be in the middle of a scene cut. Simulation results show that 98% accuracy is achieved on four half hour broadcast news [Liu98].

Anchor Identification

We mentioned that the presence of the anchorperson is important for recovering the structure of broadcast news. Liu and Huang proposed a method to adaptively detect an unspecified anchorperson in [Liu00]. As illustrated in Fig. 9.6, there are two main parts in this scheme. One is visual based detection (shown at the top) and the other is integrated audio/visual based detection. The former serves as a mechanism for initial on-line training data collection where possible anchor video frames are identified by assuming that the personal appearance (excluding the background) of the anchor remains constant within the same program.

Two different methods of visual based detection are described in this diagram. One is along the right column where audio cues are first exploited that identify the theme music segment of the given news program. From that, an anchor frame can be reliably located, from which a feature block is extracted to build an on-line visual model for the anchor. Figure 9.6 illustrates the feature blocks for two anchor frames. From this figure, it is obvious that the feature blocks capture both the style and the color of the clothes and they are independent of the image background as well as the location of the anchor. By properly scaling the features extracted from such blocks, the online anchor visual model built from such features are invariant to location, size, scale, and background. The model is used to identify all other anchor frames by a matching operation.

The other method for visual based anchor detection is used when there are no acoustic cues such as theme music present so that no image of the anchor can be reliably identified to build an online visual model. Face detection is applied and then feature blocks are identified in a similar fashion for every detected human face. Once invariant features are extracted from all of the feature blocks, dissimilarity measures are computed among all possible pair of detected persons. An agglomerative hierarchical clustering is applied to group faces into clusters that possess similar features (which indicate the same cloth with similar colors). Given the nature of the anchor's function, it is clear that the largest cluster with the most scattered appearance time corresponds to the anchor class. The use of both of the above described methods enables adaptive anchor detection in the visual domain.

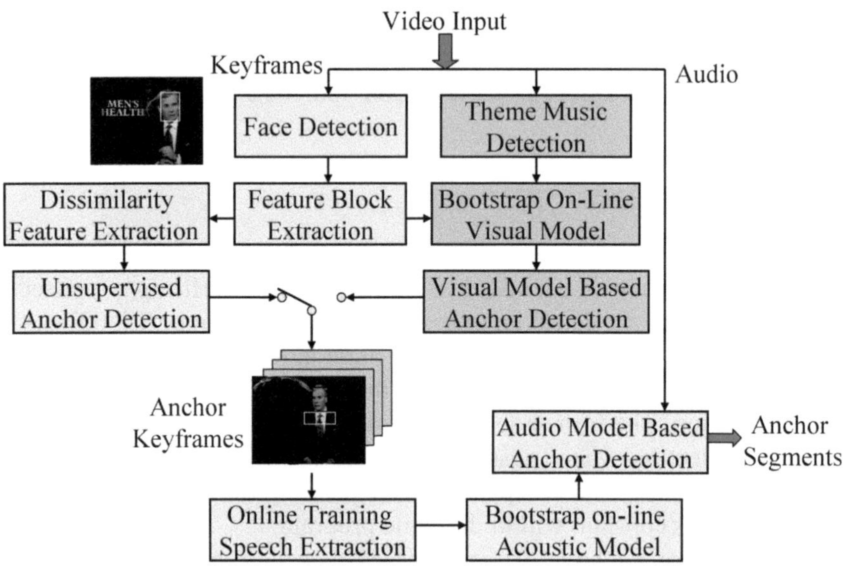

Fig. 9.6. Diagram of adaptive anchor detection algorithm.

Using only visual based anchor detection is not adequate because there are situations where the anchor speech is present but the anchor does not appear. To precisely identify all anchor segments, it is necessary to recover these segments as well. This is achieved by incorporating audio based anchor detection. The visually detected anchor keyframes from the video stream identify the locations of the anchor speech in the audio stream. Acoustic data at these locations can be gathered as the training data to build an online speaker model for the anchor, which can then be applied, together with the visual detection results, to extract all the segments from the given video where the anchor is present.

Simulation results show that the adaptive anchor detection method achieves similar performance to methods that use an offline speaker model, yet it is obvious that the adaptive methods have the full flexibility of detecting arbitrary anchors while the off-line approach does not.

News Story Extraction

Text-based discourse segmentation involves tokenization (the division of the input text into individual lexical units), grouping of processing units (granularity), similarity determination (lexical similarity between two blocks of text), and boundary identification (detection of significant lexical difference based on similarity scores). Both similarity criteria and grouping criteria affect the performance and the precision in discourse segmentation. Most work in the literature uses windows of pre-defined, fixed size for the grouping. The dilemma is that too small a window size will make similarity comparison less effective and that too large a window size can dramatically reduce the accuracy of identified boundaries. Huang et al. [Huang99] proposed a grouping criterion based on audio cues. Since anchor-based segmentation has grouped the text input into blocks, in effect, (1) adaptive granularity can be achieved that is directly related to the content, (2) the hypothesized boundaries are more natural than those obtained using a fixed window, (3) blocks formed in this way not only contain enough information for similarity comparison but also have natural breaks of chains of repeated words if true boundaries are present, (4) the original task of discourse segmentation is achieved by boundary verification, and (5) once a boundary is verified, its location is precise. This grouping scheme of integrating audio based analysis provides an excellent starting point for the similarity analysis and boundary detection.

For news content, we may organize blocks of text into four classes: news stories, story introduction, augmented news stories, and news summary of the day. The input data for text analysis is two sets of blocks of text: $T_1 = \{T_1^1, \ldots, T_1^i, \ldots, T_1^m\}$ where each T_1^k, $1 \le k \le m$, begins with the anchor person's speech; and $T_2 = \{T_2^1, \ldots, T_2^j, \ldots, T_2^n\}$ where each T_2^k, $1 \le k \le n$, contains only the anchor's speech. The blocks in both sets are all time stamped so that $T_2^k \subseteq T_1^k$. To find story boundaries, the similarity sim() between every pair (T_{b1}, T_{b2}) of adjacent blocks is computed by

$$sim(T_{b1}, T_{b2}) = \frac{\sum_w f_{w,b1} \times f_{w,b2}}{\sum_w f_{w,b1}^2 \times \sum_w f_{w,b2}^2} \qquad (9.2)$$

Here, w enumerates all the token words in each text block; $f_{w,bi}$ is the frequency of word w in block b_i, i = 1, 2; and $0 \leq \text{sim}() \leq 1$. With this approach, blocks that have a higher frequency of identical words are defined as being more similar. A threshold is experimentally set up to determine the story boundaries. After stories are segmented, set T_2 and the stories are taken as input to further extract other classes. For each story, the algorithm extracts its introduction by finding a T_2^k that has the highest similarity to that story (T_2^k does not necessarily consist of contiguous segments). An augmented story is formed by merging each story with its introduction. The news summary of the day is extracted with the criterion that it has to provide the minimum coverage for all the stories reported on that day. Therefore, it is a set of T_2^ks that together covers all the stories of the day without overlap (i.e., each story has to be introduced, but only once). With such a higher level of abstraction, users can browse desired information in a very compact form without missing the primary content.

9.2.3 Major Cast Detection

Finding the primary set of actors, or major cast, of a program is an important step for recovering the structure of the program. Here we describe the major cast detection algorithm that Liu and Wang proposed (Fig. 9.7) [Liu07]. Each major cast member is characterized by two attributes: face and speech. The detection procedure consists of finding corresponding face occurrences and speech segments by analyzing video at two levels. Audio and visual information is utilized separately at a low level, and at a higher level where cues from different modalities are combined.

At low level, the video sequence is segmented independently in both audio and visual tracks. In the audio track, clean speech chunks are extracted, within which speaker boundaries are then identified. On the other hand, the visual track is segmented into homogeneous shots, and face detection and tracking are applied within each shot. At high level, both audio and visual information are exploited based on temporal correlation among different faces and speakers. All speaker segments and face tracks are grouped using an integrated clustering method such that segments containing the same speaker and tracks consisting of the same face are merged. A list of major cast members is then constructed by associating faces and speakers to certain characters. The order of the list reflects the importance of each character, which is determined based on corresponding accumulative temporal and spatial presence.

Besides speech signals, there are other kinds of sound in audio track, for example, music, speech with music, noise, speech with noise, etc. To separate and compare different speakers, it is preferred to extract speaker information based on clean speech only. Therefore the speaker segmentation algorithm includes two steps: (1) Extract the clean speech chunks from the audio track, and (2) locate the speaker boundaries in the clean speech audio chunks.

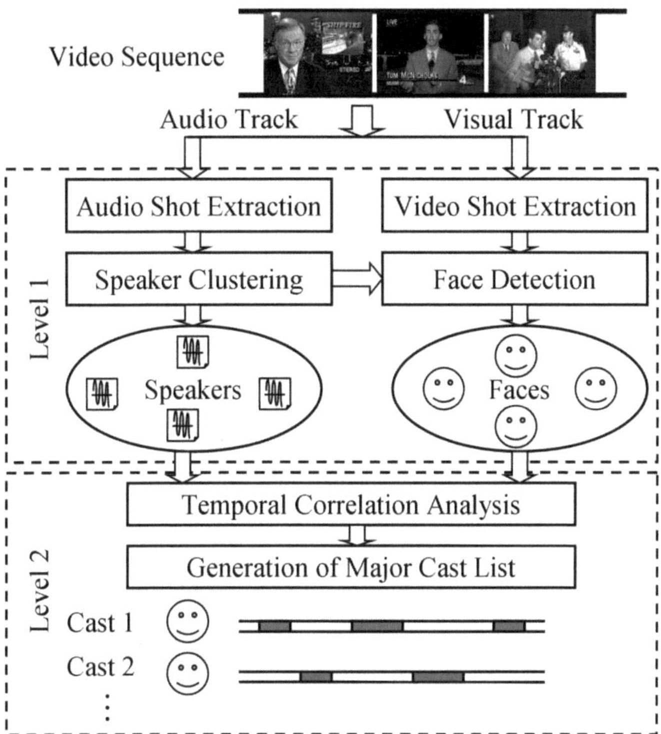

Fig. 9.7. Major cast detection algorithm.

In Chap. 6, we introduced a few face detection algorithms. Instead of tracking faces directly on the entire video, the video sequence is segmented into shots, and faces are tracked in each shot independently. Two stages are involved for face tracking within each shot: (1) detecting frontal faces in all frames, and (2) expanding face tracks in surrounding frames. In the first stage, an average face model is used to detect faces in each frame, where only frontal faces can be effectively detected. In the second stage, the detected faces are used as new face templates to search faces in neighboring frames bi-directionally. By using a detected frontal face as the template, the system can usually detect slightly tilted/turned faces of the same person, which are typically missed in the first stage. Clustering algorithms can be used to merge the face tracks of same person in different shots.

[Liu07] only considered detection of major cast appearances that are accompanied by both speech and face. Satoh et al. used visual and text information to associate faces with names [Satoh99]. The approach here is to associate faces with speech for major cast members based on the temporal correlation between faces and speakers. Following is the definition of the speaker face correlation matrix. The integrated speaker segment and face track clustering algorithm, as well as the major cast selection and ordering method, are based on this matrix.

Suppose there are M speaker segments, $S_1, S_2, ..., S_M$, and N face tracks, $F_1, F_2, ..., F_N$. Different speaker segments or face tracks may correspond to the same person. Let's assume that speaker segment S_i has L_i discontinuous sub-segments: $s^i_1, s^i_2, ..., s^i_{Li}$, each sub-segment has two attributes: starting time (ST) and ending time (ET). Similarly, face track F_i has l_i discontinuous sub-tracks: $f^i_1, f^i_2, ..., f^i_{li}$, each sub-track has three attributes: starting time, ending time, and face size (FS). Here the representative face of each face sub-track is used to determine the face size. Then the speaker face correlation (C_{SF}) matrix is an $N{\times}M$ matrix, whose element $C_{SF}(i, j)$ is defined as

$$C_{SF}(i,j) = \sum_{m=1}^{L_i}\sum_{n=1}^{l_j} OL(s^i_m, f^j_n) \times FS(f^j_n) \tag{9.3}$$

where $OL(x, y)$ is the overlapping duration of speaker sub-segment x and face sub-track y, and $FS(y)$ is the face size of y.

Figure 9.8 illustrates the correlation between speaker segment S_i and face track F_j. This definition not only considers the temporal overlap among speaker segments and face tracks, but also takes into account the effect of face size. The consideration of face size is helpful when more than one face shows up during a speech segment, where the face with the bigger size is more likely to be the real speaker.

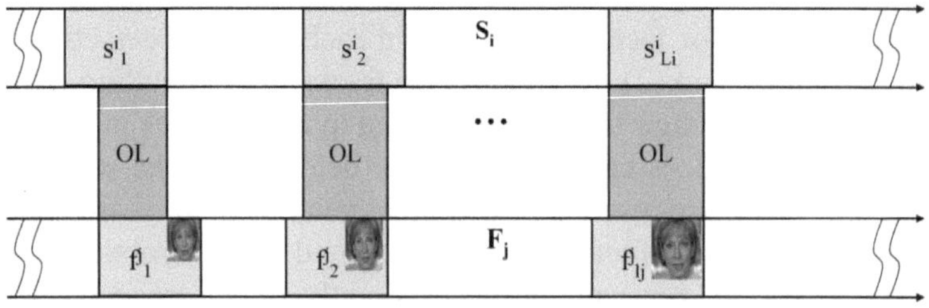

Fig. 9.8. Illustration of speaker face correlation.

The major cast is determined by linking the faces to corresponding speakers. Then, an importance score is assigned to each major cast member, so that a list of sorted major cast members is extracted. In the proposed solution, association of faces to speakers entirely depends on the speaker face correlation matrix. The value of speaker face correlation reflects both the temporal (time span) and the spatial (face size) importance of the major cast. In the following algorithm, we perform the speaker-face association and major cast ordering at the same time. Suppose there are M different speakers and N different faces, and an $M \times N$ speaker face correlation matrix C_{SF}. The algorithm is as follows:

1) Set $i = 1$.

2) Find an entry in the C_{SF} matrix with maximum value, denote the row and column indices of this entry by s_i and f_i, respectively.

3) Assign the speaker corresponding to row s_i and the face corresponding to column f_i to major cast i.

4) Remove row s_i and column f_i in C_{SF}.

5) Set $i = i + 1$, and go to step 2 unless the maximum value in C_{SF} is smaller than a threshold.

This algorithm produces a list of major cast members with corresponding correlation values, which are used as temporal-spatial importance scores. The score for each cast member essentially measures the cumulative spatial and temporal presence of this cast member.

9.3 Conclusion

Multimedia content is composed of a combination of audio, video, text, images, animation, etc. Information carried in multimedia data is distributed across all constituent parts. To effectively analyze the multimedia content, we need to process all available media and fuse the knowledge together. This chapter described a few multimedia content processing techniques using three examples: closed caption alignment, multimodal content segmentation, and major cast detection. Through these three cases, we show the advantage of multimedia processing, compared to isolated text, audio, or image processing.

References

[Adams03] Adams, W.H., Iyengar, G., Lin, C-Y, Naphade, M.R., Neti, C., Nock, H.J., Smith, J.R.: Semantic indexing of multimedia content using visual, audio and text cues. *Eurasip Journal on Applied Signal Processing*, **2** (2003).

[Boreczky98] Boreczky, J.S. and Wilcox, L.D.: A hidden Markov model framework for video segmentation using audio and image features. *ICASSP*, **6**, pp. 3741–3744 (1998).

[Chen98] Chen, T. and Rao, R.R.: Audio-visual integration in multimodal communication. *Proceedings of the IEEE*, **86**(5), pp. 837–852 (1998).

[Christel05] Christel, M. and Conescu, R.: Addressing the challenge of visual information access from digital image and video libraries. *JCDL* (2005).

[Cutler00] Cutler, R. and Davis, L.: Look who's talking: Speaker detection using video and audio correlation. *ICME* (2000).

[Fisher00] Fisher, J.W., Darrell, T., Freeman, W.T., Viola, P.: Learning joint statistical models for audio-visual fusion and segregation. Advances in Neural Information Processing Systems (2000).

[Gibbon06] Gibbon, D., Liu, Z., Shahraray, B.: The MIRACLE video search engine. *IEEE CCNC* (2006).

[Huang98] Huang, J., Liu, Z., Wang, Y.: Integration of audio and visual information for content-based video segmentation. *ICIP*, **3**, pp. 526–530 (1998).

[Huang99] Huang, Q., Liu, Z., Rosenberg, A., Gibbon, D., Shahraray, B.: Automated semantic structure reconstruction and representation generation for broadcast news. *SPIE* (1999).

[Li03] Li, Y. and Kuo, C.: *Video content analysis using multimodal information: For movie content extraction, indexing and representation*. Springer (2003).

[LiD03] Li, D., Dimitrova, N., Li, M., Sethi, I.: Multimedia content processing through cross-modal association. *ACM Multimedia* (2003).

[Lienhart99] Lienhart, R., Pfeiffer, S., Effelsberg, W.: Scene determination based on video and audio features. *IEEE Int. Conf. Multimedia Computing and Systems*, **1**, pp. 685–690 (1999).

[Liu98] Liu, Z. and Huang, Q.: Classification of audio events in broadcast news. IEEE Signal Processing Society Workshop on Multimedia Processing, pp. 364–369 (1998).

[Liu00] Liu, Z. and Huang, Q.: Adaptive anchor detection using online trained audio/visual model. *Proc. of SPIE* (2000).

[Liu06] Liu, Z., Gibbon, D., Zavesky, E., Shahraray, B., Haffner, P.: AT&T research at TRECVID 2006. TRECVID 2006 Work-shop (2006).

[Liu07] Liu, Z. and Wang, Y.: Major cast detection in video using both speaker and face Information. *IEEE Transaction on Multimedia*, **9**(1), pp. 89 – 101 (2007).

[Marvel07] MARVEL: Multimedia analysis and retrieval system. Intelligent Information Management Dept., IBM T. J. Watson Research Center, http://mp7.watson.ibm.com/marvel/, cited 10 Dec 2007.

[Saraceno98] Saraceno, C. and Leonardi, R.: Identification of story units in audio-visual sequences by joint audio and video processing. *ICIP*, **1**, pp. 363–367 (1998).

[Satoh99] Satoh, S., Nakamura, Y., Kanade, T.: Name-it: Naming and detecting faces in news videos. *IEEE Multimedia Magazine*, **6**(1), pp. 22–35 (1999).

[Snoek05] Snoek, C., and Worring, M.: Multimodal video indexing: A review of the state-of-the-art. *Multimedia Tools and Applications*, 25(1), pp. 5–35 (2005).

[Wang00] Wang, Y., Liu, Z., Huang, J.: Multimedia content analysis using audio and visual information. *IEEE Signal Processing Magazine*, **17**(6), pp. 12-36 (2000).

10 Research Systems

10.1 Introduction

Over the years there have been numerous research contributions to the field of video and audio retrieval from the academic community as well as rapid innovation on the commercial side from, companies including Internet startups. Database systems vendors have incorporated algorithmic advances via modular architectures (e.g. Informix® Datablades, Oracle® Cartridges). Given the sheer number of deployments of video search systems in industry and on the Web, we will focus on a subset that are of particular technical interest, in order to illustrate the concepts presented in this book. An exhaustive treatment is beyond our scope, and interested readers should consult the valuable surveys of this field provided in [Hanjalic04, Lew06, Tjondronegoro07]. We will focus in greater detail on a particular system to provide an end-to-end practical perspective on video search. Table 10.1 Table 10.1 lists some of the systems mentioned in this chapter.

Table 10.1. Representative multimedia retrieval systems.

Name	Organization	Year Initiated
VMR – Video Mail Retrieval	Cambridge	1994
Informedia	CMU	1994
MIRACLE	AT&T	1994
QBIC	IBM	1995
VideoLogger	Virage	1996
WebSeek	Columbia Univ.	1996
BNN – Broadcast News Navigator	MITRE	1997
VideoQ	Columbia Univ.	1998
SpeechBot	Compaq / HP	1999
TALES	IBM	2006

10.2 Academic and Industrial Research

One of the first groups to propose exploiting closed captions for relevant media retrieval was the MIT Media Lab. Their Network Plus system [Bender88] addressed the issue of personalization by filtering content from a broad spectrum of broadcast or published sources based on specified user interests. This is a topic of great interest today as more and more media is being produced and current technologies make implementing this vision much more practical. The Media Lab also introduced "Salient Stills" which are single-frame visual summaries that are appropriate for certain shots such as pan and zoom with fixed backgrounds [Teodosio93] where optical flow is used to compute a composite image.

Cambridge University's Video mail retrieval (VMR) project was very much ahead of its time in that it leveraged a novel asynchronous transfer mode (ATM) local video network [Jones94] as well as applied information retrieval (IR) methods with word spotting to enable true automated content-based retrieval of video media. The system also supported retrieval using the metadata that one would expect with a messaging system (sender name, time/date, duration, etc.) The team was one of the first to study IR in the presence of ASR error and conclude a perhaps counterintuitive result that has been later confirmed in different contexts on several occasions that the performance is only slightly impaired as compared with IR using manual transcriptions [Jones95].

The Informedia project at Carnegie Mellon University (CMU) which began in 1994 was sponsored by the NSF Digital Libraries Initiative and its descendant projects are still active at CMU [Cristel95, Wactlar96]. An early instantiation ingested MPEG-1 video and used speech recognition (CMU's well known Sphinx-II system) for news retrieval and supported speech queries [Hauptman95]. Video processing included an impressive suite of techniques including shot boundary detection, representative frame selection, face detection and optical character recognition (OCR). Closed captions were used when available with alignment to improve synchronization and multimodal video segmentation created visual representations used in interactive visual interfaces. Later, the concept of a "Video Skim" created visual summaries or shortened versions of longer video sequences via multimodal processing that attempted to convey the pertinent information [Smith96]. By observing that closed captioned television represents a vast resource of partially labeled data, researchers created an example of a data-driven system that could effectively learn over time. They showed how this resource could be tapped for adapting language models [Jang99]. "Name-it" was a multimodal processing approach for detecting and identifying persons in video [Satoh99]. Entity extraction, combined with geographic information processing was explored in the context of news information retrieval and browsing using spatial constraints [Christel00]. Recent work includes extreme interfaces where the advanced graphics capabilities and system throughput of today's personal computers are employed to enable users to rapidly browse and interact with volumes of rich media content [Worring07]. CMU commercialized their indexing technology under the name MediaSite.

Columbia University's Digital Video Multimedia Laboratory (DVMM) has made numerous contributions to this field. Projects such as WebSeek, VisualSeek [Smith96], VideoQ [Chang98] embody the vision of creating novel media processing and retrieval techniques and demonstrating them on real-world data with interactive response times for user queries. VideoQ goes beyond image based retrieval to enable true video search using motion queries enabled by video object segmentation and tracking. The group has been very active in TRECVID evaluations and has developed the Columbia374 dataset for high-level concept detection benchmarking (see below). In addition to visual semantic classification and search systems, the DVMM lab has projects on video summarization, mining, near duplicate detection.

IBM's heritage of developing information management technology in general is well known, and the area of multimedia is no exception. An active research program in multimedia information retrieval has been ongoing for a number of years. IBM's groundbreaking Query By Image Content (QBIC™) system [Flickner99] is well known as one of the first successful projects aimed at content based image retrieval. The CueVideo work employed automated processing for shot boundary detection and allowed for visual browsing of indexed video content. IBM has contributed to the MPEG-7 standardization efforts, and has proven MPEG-7's effectiveness by creating and deploying tools and end-to-end systems based on the technology (e.g., Marvel). Their Unstructured Information Management Architecture (UIMA) is intended for multimedia search applications such as those that we have been describing. Their Translingual Automatic Language Exploitation System (TALES) supports advanced multilingual video analysis including speech to text and real-time translation with cross-lingual search. The system has been instantiated and deployed for continuous operation with four video feeds in a single rack-mount configuration for Arabic and Chinese news monitoring applications [Roukos06].

MITRE, a non-profit federally funded research and technology organization, developed the Broadcast News Navigator (BNN) in 1997 [maybery97] which included particular emphasis on multimodal story segmentation. More recently they have focused on fine-grained content personalization and created the P-BNN system [maybery04] which effectively produces "personalcasts" based on user interest profiling. Metadata is generated for content segments beyond what is supplied with electronic program guides. The system includes query expansion and refinement along with relevance feedback and local context analysis for tailoring the source content to match the users' interests.

AT&T's Bell Labs (and later AT&T Labs) has made numerous contributions to telecommunications technology, from fundamental information theory up through speech processing and video compression standards such as MPEG-4. In the early 1990s, the Machine Perception Research department began developing real-time video segmentation algorithms that went beyond shot-boundary detection to include intra-shot sampling based on analyzing the camera operations [Shahrary95]. Unlike other algorithms, the method relied primarily on motion features rather than color histogram features. Efficient processing enabled these more complex features to be computed in a practical system. Based on this video segmentation, the team experimented with techniques for selecting compact sets of representative images that would best convey the visual content of the video. Additional media processing including natural language processing of closed captions was used to build 'condensed' versions of broadcast television programs. These very low bitrate representations were ideal for printing and network delivery to subscribers using the technology available at the time which consisted of dial-up modems that were incapable of delivering high-quality full-motion video. While replaying representative images along with text and optional audio streams achieved the goal of high media compression, it still required isochronous playback, which had two negative implications: (1) although comparatively low, there is still a minimum network bandwidth required to deliver the content, and (2) the viewer could not easily consume the content faster than the original media presentation duration. To address these concerns, AT&T proposed producing printed versions of video programs called "Pictorial Transcripts" [Shahraray95a]. These compact representations of video programs were ideally suited for delivery to Web browsers which were emerging at the time since they effectively converted video programs into Web documents.

While originally developed to overcome the limitations of networking and terminal device technology of the day (this was an early example of content adaptation), the aspect that these representations could be viewed faster than real-time provided enduring redeeming value. As we have seen, browsing results sets is a key component for video search engine systems. By combining the lightweight video summaries available in the form of pictorial transcripts with video archiving and streaming services, systems were constructed that enabled rapid access to large video collections in a manner that seamlessly leveraged available networking and computing resources. For example, broadband-connected users could rapidly page through dynamically generated video results formatted as documents and then use these to interact with and control high-quality streaming video playback. Users with less capable devices could view page representations which captured the essence of the video content.

The AT&T research group focused mainly on distributed architectures for delivery of video and audio material using content-based indexing, which matched well with a telecommunications company's business interests. Today these architectures are the norm given the advances in IP media delivery. These concepts were embodied in their Digital Video Library (DVL) which supported rapid access to large video archives streamed over the Internet. To extend the application areas beyond closed captioned or subtitled content, AT&T drew from decades of speech processing research and employed large vocabulary automatic speech recognition (LVASR) based on their Watson recognizer. While speech was used for retrieval, parallel text alignment methods were employed to create very high quality multimedia documents using post-production scripts or manually prepared transcriptions. The team developed media personalization and device adaptation methods to facilitate "lean back" content access [Gibbon03]. For telephony applications where acoustic conditions are relatively poor, AT&T investigated phonetic and lattice search techniques [Saraclar04]. More recently, the group developed the MIRACLE video search engine which incorporated EPG metadata from DVR sources as well as other content sources such as Podcasts [Gibbon06, Liu06].

In 1997 AT&T participated in the TREC Spoken Document Retrieval and incorporated document expansion to improve performance when ASR conditions are poor [Singhal97]. The SCANMail project developed technologies including entity detection, customized language models and advanced user interfaces for voice mail applications [Whittaker02]. The PRISM (portal infrastructure for streaming media) project focused on efficient delivery and resource identification schemes for broadcast television content [Basso00] and SBTV (searchable browsable TV) investigated content indexing for RTSP delivery of MPEG-2 media [Gibbon99]. At the AT&T Cambridge Research Labs, the AT&TV project built on the Digital Asset Retrieval Technology project (DART), which focused primarily on personal media collections, to support continuous acquisition of multiple broadcast channels stored in MPEG-1 format [Mills00]. Recently AT&T

has participated in the TRECVID shot boundary detection evaluations and collaborated with Columbia University on other tasks [Liu07].

Many other enterprises whose products and services relate to media, entertainment or consumer electronics have active research groups that continue to provide results for the community at large. The computing industry, e.g. Microsoft Research, Intel, etc. have also played an important role in technology development since handling media is a key capability of today's computing platforms. Many exploratory systems have been developed that bring media content processing and retrieval methods to bear on applications of interest to the constituent entities. For example, at Phillips, the Video Scout system demonstrated advanced media personalization capabilities for DVR applications [Demetrova03].

10.3 Early Internet Deployments

As we saw in Chap. 1, there is a wide range of Web sites offering some form of video search. Here we focus on some of the early Internet deployments and in particular, sites with a focus on automated media analysis for retrieval. Again, this is not an exhaustive list, but some representative and well known examples are provided.

10.3.1 SpeechBot

SpeechBot was developed at the Compaq (later HP) Cambridge Research Labs in 1999 and was one of the first speech indexing systems to ingest large volumes of Internet content. "By early June 2003, the seven-member group, based at HP's Cambridge Research Laboratory in Cambridge, Mass. (USA) had catalogued more than 17,000 hours of multimedia content – making SpeechBot the largest multimedia index in the world" [Stuart03]. As RSS with media enclosures was not in widespread use at the time, SpeechBot used a traditional crawler architecture to obtain media files for indexing.

SpeechBot used the Calista recognizer based on HMMs [VanThong00] trained using the HUB-4 1998 training corpus, and had a 64K word vocabulary with 4M bigrams and 15M trigrams. The word error rate was reported to be 20% for studio content, up to 50% for lower quality speech.

Interestingly, the acoustic models were generated by compressing the training data to form a better match to the content typically encountered on the Web at the time (Real audio was prevalent). Also, the sampling rate selected was 8 kHz which is generally associated only with telephony bandwidth speech today. Processing took 6 to 30 times real-time using 450 MHz Pentium II processors. SpeechBot segmented long-form content into fixed-length units (20 seconds) to help the IR engine identify revenant content and metadata was also added to the index [Eberman99]. The team also built BoogeBot for music search which finds similar songs to a given sample in a database of 18,000 songs. SpeechBot went offline in 2005 and is now no longer available.

10.3.2 StreamSage

StreamSage was founded in 2000 and acquired by Comcast in 2005. It provides video search services for Comcast's broadband subscribers (videosearch.comcast.net). Videos are displayed using a unique video browsing user interface based on flash and a circular thumbnail display ("The Fan"). StreamSage has participated in TRECVID evaluations [Rennert03] and has developed systems using term mutual information and topic segmentation to improve retrieval performance [Davis04].

10.3.3 SingingFish

Around the height of the dot com era, SingingFish emerged and enjoyed great success as the media search engine tied into the media players of the two dominant streaming media systems of the time: Real and Windows-Media. Founded in 1999 and later acquired by Thompson and then AOL, SingingFish employed its Asterias crawler to discover content and it also developed a content producer program for managed content contribution. This hybrid approach to content acquisition is common today as most major search engines include an aggregation aspect. Many Internet users are not familiar with the name, since SingingFish typically provided search results through third party end user media applications or Web media destinations / portals [Fritz03]. SingingFish has employed MPEG-7 and made contributions to using MPEG-7 for media search, including extensions to the Media Format description scheme to improve efficiency [Rehm00].

10.4 Selected Commercial Systems

One of the most widely known video search companies, Virage, grew out of the observation that image analysis must be domain-specific to be successful, and yet a common infrastructure or architecture can be defined for a wide range of applications [Bach06]. This philosophy is extended in the context of video retrieval as well as image retrieval. Virage produced several commercial systems including VideoLogger which supported video segmentation, closed caption and speech retrieval among other media analysis modules in a "pluggable" architecture, with a flexible file format for storing the indexing results (called VDF).

10.4.1 Virage and Convera

Virage faced competition from other startups as well as more established information retrieval product and service suppliers such as Excalibur (now called Convera). Many of these companies developed systems for specific applications or data collections and in many cases these data collections were evolving from text to include multimedia content as well. Convera developed a product called ScreeningRoom that incorporated advanced media analysis and supported SQL for integration with databases such as Oracle. Media asset management systems (e.g. Artesia) and production tools (e.g. Avid) vendors also saw the value of automated media indexing for many production and archival management applications and sometimes formed alliances or otherwise provided some measure of support for media logging applications such as Virage's VideoLogger.

10.4.2 Nexidia (FastTalk)

Nexidia (formerly FastTalk) supports a wide range of audio search applications using phonetic search. They create a "phonetic audio track" or .PAT file for each asset [Nexedia08]. The phonetic approach circumvents the out-of-vocabulary problem and facilitates supporting multiple languages without requiring custom dictionaries. The company has close ties with the Georgia Institute of Technology. Fast-Talk was "founded in 2000 based upon basic research at Georgia Tech's Interactive Media Technology Center" [FastTalk08]. At the time, the focus was on word-spotting but later expanded to include phonetic search, transcription alignment, and other audio tools.

10.5 Resources: Datasets, Evaluations, Conferences

Collections of multimedia data aid researchers in algorithm development and can be indicative of the progress of the state of the art over time. For example, in the image coding community, one can track the reconstructed image quality for a given bitrate over the years as new coding algorithms and standards were developed. Similarly for speech recognition tasks, the word error rate, perhaps at a real-time operating point, can be studied for different systems. Unfortunately, the video information retrieval community has not enjoyed the same level of availability of reference datasets as has been the case for speech and text data. For most organizations responsible for producing video, their archival collections are viewed as strategic assets, and releasing them even under strict licensing agreements is deemed not to be worth the risk of potential loss. Video sources in the public domain such as archive.org may not be a good match for the applications under study due to the age or genres of the content. In Chap. 2 we presented application areas and sources of data for which video search engine systems have been built; here we focus on datasets that have been labeled with ground truth for algorithm development.

Table 10.2 lists a handful of datasets that have been used by the multimedia information retrieval researchers along with the responsible organizations. Where an organization has multiple similar datasets, a single representative is chosen for brevity. The TRECVID program each year produces datasets for laboratory style evaluations focused specifically on video information retrieval applications. For uni-modal processing tasks such as face recognition and speaker identification, other organizations maintain datasets and these are used for multimedia applications as well. The Linguistic Data Consortium provides a broad range of datasets for speech and natural language research, and also works with NIST to provide TRECVID data. The HUB-4 set was widely used by the multimedia information retrieval community for systems focused in the broadcast news domain. The Disruptive Technology Office (DT) has developed the Large Scale Concept Ontology for Multimedia (LSCOM) and Columbia University and CMU have annotated TRECVID 05/06 keyframes based on a selected subset of concepts. Along with this, Columbia University provides extracted low level image features used to train models, as well as the labels obtained via late fusion of several classifiers to serve as an accuracy benchmark. This dataset, known as the Columbia374 [Yana07], is a selected subset of 374 high level concepts and has begun to be used by other groups (Viero-374). Earlier, the MediaMill Challenge set [Snoek06] identified 101 semantic concepts.

Table 10.2. A sampling of multimedia retrieval datasets.

Dataset	Media	Organization	Comments
TRECVID	Video/Audio, keyframes, transcriptions	NIST / LDC	Multiple datasets; new data added each year
MediaMill Challenge	Annotated keyframes	MediaMill	101 semantic concepts base on the TRECVID 05/06 dataset
Columbia374	Annotated keyframes	Columbia University	374 semantic concepts base on the TRECVID 05/06 dataset
HUB-4	Speech, Transcriptions	LDC	Broadcast News ~1996
TDT	Text	LDC	Reuters and CNN news stories used by MM researchers for topic segmentation
EARS/MDE	Speech, Annotations	LDC	Metadata extraction from speech.
FERET	Images	NIST	Face Recognition, ~1993
FRGC	Images, 3D data	NIST, et al.	Face Recognition Grand Challenge, High resolution, 3D, multi-view
Corel	Images		Image Retrieval
VACE		ARDA	Video Analysis and Content Extraction
Wordnet	Text	Princeton University	Synonyms, relations, for NLP work
Penn Treebank	Text	U. Penn	linguistic structure, tagging

The Linguistic Data Consortium (LDC) provides an invaluable function for natural language research. Their datasets are the benchmark for many speech recognition and natural language processing tasks. Recently, keeping in sync with current research trends, the nature of the datasets has evolved from a focus on speech to text (STT) to rich transcriptions and metadata extraction. One of the goals of this is to "enable technology that can take the raw STT output and refine it into forms that are of more use to humans and to downstream automatic processes." [Strassel03].

In addition to these valuable datasets for benchmarking algorithm performance, many groups make baseline implementations of algorithms available. Also, there is a wide range of applicable tools available to be used as building blocks for media processing algorithm development (e.g. Intel's OpenCV Open Source Computer Vision Library).

10.6 Media Monitoring Deployments

Broadcast monitoring services employ video search at large scale and with high reliability, and yet most Internet users are unaware that they exist. These services capture broadcast television content including local programming continuously and in different geographic regions. They create and maintain databases that support keyword search on the closed caption and provide additional services such as alerting. The services can incorporate data feeds from Nielson Media Research™ to indicate the audience size and use SQAD™ to provide information about advertisement costs. For example, Video Monitoring Services (VMS™), founded in 1981, monitors all 210 defined metropolitan areas (DMAs) in the US and can provide near real-time Web access to the captured media [VMS08]. Critical Mention™ was founded in 2002 and provides broadcast monitoring services for a wide range of applications. They maintain a database of over 5 million "clips" and "25 TB of indexed television content" [Critical08]. Incidentally, media monitoring in one form or another has been existence for many years, for example BurrellesLuce has been in the business for over a century [Burrelles08]. The TALES system mentioned above is intended for media monitoring applications as well, primarily for multilingual applications. A suite of services related to broadcast monitoring and offered by some of the same companies is based around maintaining advertisement databases. These services extend the range of applications beyond news alerting to include competitive ad monitoring for enterprise customers.

10.7 Case Study: AT&T MIRACLE

10.7.1 Introduction

In order to give readers an intuition for some of the practical issues encountered in building content based retrieval systems, we will take a look at the MIRACLE (Multimedia Information Retrieval by Content) system in detail. This system has been developed over a number of years and is maintained by AT&T researchers as a platform for media processing algorithm development. The system ingests content on a daily basis from broadcast TV as well as Internet sources. Both continuous ingest and EPG driven acquisition modes for selected broadcast programs are supported.

The design goals include efficiency and ease of maintenance, but above all, flexibility and extensibility. There is an archival aspect which implies that older content may have been processed with earlier versions of algorithms, or may not have been subjected to newly developed media processing modalities at all. Similarly, content is ingested from a variety of sources with varying degrees of available source metadata. And finally, the applications supported by the platform run the gamut from personalized mobile media through video data mining for trained users with workstations, to advanced content processing for IPTV service prototyping. This requires extensibility and well defined system interfaces (e.g. Web services.)

10.7.2 System Architecture

The overall architecture follows the framework that we have been discussing, a content acquisition block, followed by content processing and then archival storage with a query processing and user interface module (Fig. 10.1), where CMS refers to a content management system. In this case, copies of the content are stored in archival servers, and in fact, multiple versions of each asset are produced via transcoding for different applications. Logically, it is easiest to think of the processing in terms of batch operations on media files, but the system is also capable of operating in a real-time mode in which the processing operations begin as the content is being acquired to minimize the end-to-end latency between content acquisition and posting to the content servers.

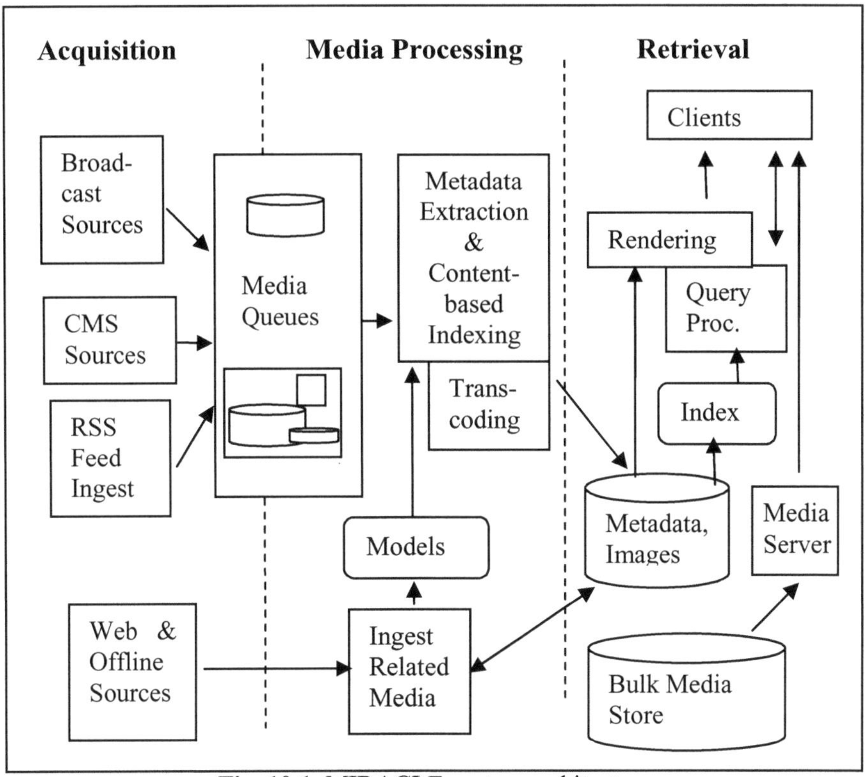

Fig. 10.1. MIRACLE system architecture.

10.7.3 Collections

The system ingests content from a variety of sources. The main collections are as follows: broadcast television sources, video and audio podcasts, and enterprise video. The average runtime for a program is about 45 minutes and the number of assets for these collections is indicated in Table 10.3. Other collections not shown include user generated content, rushes material with annotations, lectures with slides, and collection of a multilingual movie subtitles in over 20 languages. Although this is a prototyping platform and the size of the collections is somewhat modest, the scale is large enough to give a reasonable indication of system performance when deployed. Note that each asset includes metadata and keyframes and typically includes multiple copies of the source media at different bitrates and in different encoding formats. Fig. 10.2 shows that over time, higher quality renditions became practical to process and maintain on-line. Additionally, the available source metadata for each asset improved over time as better sources such as EPG and RSS were added to the platform. It is the responsibility of processing algorithms and user interfaces built for MIRACLE to exploit available metadata and media, but gracefully cope with the fact that the database contains assets with different amounts of metadata, and that there are various media representations available for a given asset (Fig. 10.2). Although the timeline in the figure ends at 2006, collection of video material continues and the capability to acquire higher bitrate HD video has been added.

Table 10.3. Metadata and size of several MIRACLE collections.

Collection	Text Source	Assets	Metadata
Broadcast TV	Closed Captions	50,000	EPG
Broadcast TV	Spanish Lang. CC	2,000	EPG
Broadcast TV	Speech Recognition	20,000	EPG
Broadcast TV	Transcripts	4,000	EPG
Video Podcasts	Speech Recognition	4,000	RSS
Audio Podcasts	Speech Recognition	5,000	RSS
Enterprise Video	Pre-Production Scripts	600	CMS

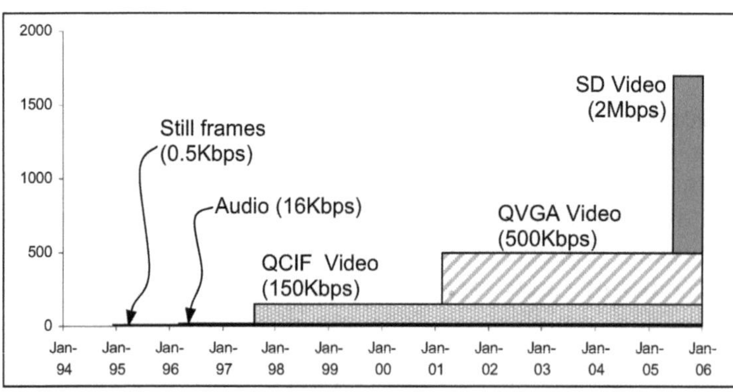

Fig. 10.2. Media bitrates for MIRACLE assets as a function of time.

10.7.4 Data Organization

Each media item (or file) ingested into the system is assigned a unique content identifier (CID), and this is mapped to a filesystem path into the archive for storage of the media and derived metadata. In addition to the CID, each asset is described by a minimum of four descriptors as follows:

1. owner identifier – entity responsible for originating the content;
2. series identifier – for episodic or recurring programs, this identifies the series;
3. series title – text string describing the series;
4. creation date – UTC indicating the asset creation date and time.

As can be seen, this data is easily derived from typical metadata sources such as EPG and RSS. However, the system is capable of synthesizing these fields in rare cases if necessary given simply a bare media file. In this case, assuming no metadata is available from the media file container, the filesystem metadata indicating file owner and creation date is used and the filename is used to populate the title and series name fields. These fields are indexed so that queries for lists of all content from a given owner or for a given series can be quickly processed.

Any additional metadata that is available such as the episode title, description, date aired, etc. is captured and maintained in XML format. For content already described by XML such as RSS or ADI, the relevant XML elements for the media item are extracted and preserved in the archive along with the asset. If required by the application and justified by the volume of assets, these additional fields can be indexed in addition to the basic attributes described above.

Each textual representation of the media item is individually maintained and indexed in the archive. Primarily these are various transcriptions of the dialog, but of course for some content there is no dialog (such as for some user generated content or raw footage), and the system and applications are designed to accommodate this. If available, field annotations are indexed (e.g. an annotation may describe a wide shot in a town as "WS of street"). The dialog representations may include one-best speech transcription, closed caption, as-aired transcriptions, teleprompter feeds, subtitles and translations.

To summarize, each media asset in the system has a base set of descriptive tags, may have additional metadata, and may have zero or more textual representations of the dialog or annotations. Further, there may be multiple versions of the linear media at different bitrates and in different encoding formats. Managing this heterogeneity increases the complexity of the system and applications, but results in increased generality, and facilitates exploratory research.

10.7.5 Acquisition / Ingest

The purpose of the acquisition modules is to locate and obtain content and metadata and to post that content to the processing module. While the system has the capability to ingest content from a wide range of sources (e.g. H.264 multicast video from professional encoders typically used in IPTV applications), we will focus on the two primary acquisition modules for MIRACLE which handle content recorded from DVRs and RSS sources. The acquisition modules record or download requested content as specified by the system configuration. They also check with the database to insure that multiple copies of the same content are not inserted.

Consumer grade DVRs don't offer the quality or reliability of broadcast quality equipment and encoders. However, there are several redeeming qualities of DVRs including:

1. Ease of configuration – designed for consumers, so EPG configuration and program scheduling are simplified and can even be done with an IR remote.
2. Acceptable video quality – digital broadcasts (ATSC DTV or DVB) can be captured directly. While this is far from archival quality, good quality transcoded versions for streaming applications can be derived. Even with analog video capture and local MPEG-2 encoding, the quality is acceptable for many applications such as creating proxy quality for Internet delivery.
3. Metadata management – EPG data and closed caption are captured.
4. Developer community – related tools are available from a large community of developers.
5. Low cost – orders of magnitude lower than corresponding professional equipment.

In one instance of the platform, multiple Windows Media Center Edition DVRs are used to gain a measure of redundancy by programming more than one system to record a particular feed. These systems maintain a buffer of typically 10 days worth of recorded content in DVR-MS format which provides an additional measure of system resilience. The recording format conveniently encapsulates MPEG-2 and EPG metadata in a single file and can support encryption. For analog acquisitions the systems are configured to encode 720x480 NTSC at 6Mb/s MPEG-2. For ATSC, other resolutions such as 704x480 SD and 1920x1080 HD resolutions can be captured at up to 9.5GB / hour. The DVRs are configured to record content in one of two ways:

1. Episodic – several programs are selected (e.g. evening news) for each DVR and for each program, each episode is recorded. Repeat broadcasts are not processed. The programs may be received on different channels and the DVR will change the channel prior to recording.
2. Continuous – the DVR is configured to record a single channel continuously, 24 hours a day, 7 days a week. All content is processed, including any repeat broadcasts.

In addition to using DVRs for content acquisition, the system also processes media files from the Web in the form of audio and video Podcasts. While this content source obviously differs from DVR acquired content in available metadata, resolution and encoding, the system attempts to normalize these disparate sources as much as possible so that applications need not be concerned with the content source. The acquisition module essentially performs the same functions as an RSS feed reader which is to periodically check a predetermined list of feeds for any new content and download the media components and metadata. A partial metadata mapping operation is performed to populate the above described minimum required fields as well as a few optional fields such as the description. In this case, the "owner" corresponds to the origin RSS URI IP name, and the "series name" is derived from the URI path. Additionally the source RSS metadata is preserved since it may contain additional metadata that may be of value for applications. This architecture offers inherent redundancy for search engines because the content is buffered at the source, and the reader can make multiple attempts to ingest the content if necessary. While some sources maintain 10 or more recent episodes, some sites only post the most recent episode, so this somewhat limits the acquisition system failure resilience. While reading RSS feeds is certainly less complex than capturing live broadcasts (for one thing, no special hardware is required), there is more variability in the content sources in terms of media resolutions and formats that the processing system must be able to manage. Also, this content does not include closed captioning, so the quality of the text representations is limited to speech recognition output. This architecture is easily extended for ingesting VoD content packages in ADI 1.1 and MPEG-21 format.

10.7.6 Content Processing

The acquisition modules push their content up to a centralized queue and a cluster of processing nodes performs the operations of metadata extraction and transcoding. For high bitrate media, an optimization is implemented to reduce network accesses by distributing this logical queue across the physical disks in the processing nodes. The processing operations are indicated by Fig. 10.3, which shows a particular flow of an individual media asset through the system. The processing modules are invoked as appropriate for the asset (e.g. closed caption processing is not invoked for Podcast content.) Each module represents its analysis results in XML and these are filtered and combined into a compilation XML representation so that applications can easily access all of the results of media processing in a single operation. Also, additional modules may be easily added as new algorithms are developed and computing resources become available. For example, for a subset of the content, semantic classifiers (i.e. Columbia374) are applied after content based sampling. Such additional processing can easily be applied to the archived material and the results captured by injecting additional elements into the XML representations. In fact such a metadata augmentation operation is preformed on a regular basis to import transcriptions that become available for some broadcast content within a day or two of the broadcasts. A process similar to an RSS feed reader downloads transcripts and performs a text alignment (similar to the closed caption alignment in Fig. 10.3) and generates a new, higher quality metadata description of the corresponding asset.

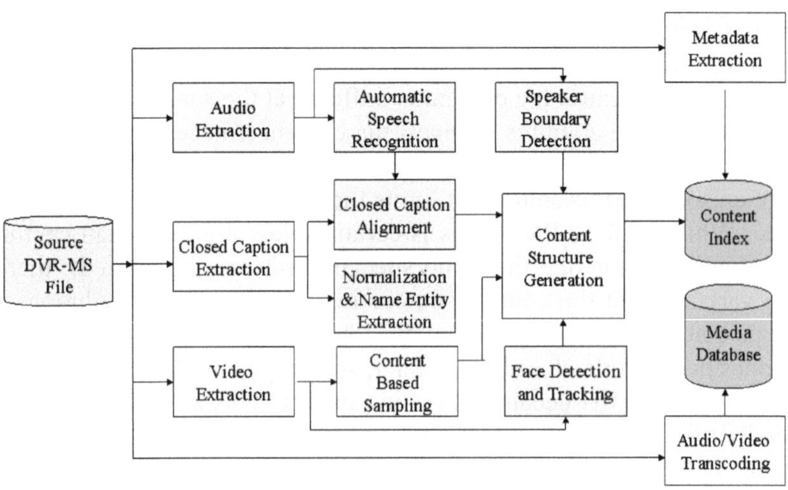

Fig. 10.3. MIRACLE content processing.

10.7.7 Real-time processing

The batch mode processing described above is designed for scalability (additional processing nodes can be added as needed) and reliability (content is buffered and can be reprocessed if necessary.) However, for applications such as generating personalized multimedia alerts of news events, end-to-end latency is a critical design parameter. Since most of the processing modules operate locally on the media stream, the system is designed to be operated in a dataflow configuration as well. This way, as the media is written to disk by the DVR, it is read by the content processing thread so that the index will be available immediately after the recording concludes. To further reduce latency, the modules are designed to stream their results to disk so that alerting or publishing modules can operate on the index while the content is being recorded. While in batch mode the only real-time requirement for the DVR is that it writes the compressed media to disk without dropping data, for real-time processing we must carefully select of a set of processing modules based on the CPU utilization of the DVR. The system is configured so that the processing operates somewhat faster than real-time on average. For modules that require offline processing (e.g. multimodal topic segmentation,) a second pass post-processing operation is performed. This real-time configuration is used for the continuous acquisition mode described above.

10.7.8 Query Engine

The media, representative frames and metadata are stored in a central repository and organized into logical collections (e.g. broadcast TV with Spanish language Closed Captions, Enterprise content with scripts, etc.) Indexing services operate on the textual content and the global metadata asynchronously to build a content index.. Additionally other databases index the XML-formatted metadata to support more advanced content queries such as visual concept queries. A two-phase process is used to render HTML results pages based on textual and global metadata (e.g. programs about NASA this week). First, a ranked set of documents matching the query is generated, this is then broken into pages of results (typically 20 items) for the user. For each page, the asset metadata such as title and date are displayed along with a "video paragraph" consisting of a thumbnail and matching query context for the most relevant segment of the program.

For different applications the relevant metadata is transformed to build user interfaces in different ways through a rendering operation. This not only allows for different layout styles, but also filters the available metadata so that only the fields pertinent to the particular application are processed. Finally, based on the application or user preferences, content locators are used to retrieve the relevant media with a range of delivery options by means of a media metafile generator. The generator builds ASX or SMIL to position long-form media at the relevant starting point. Alternatively, media player plug-ins can be used with client side scripting for media positioning.

10.7.9 Applications

The system supports an expanding set of applications ranging from a simplified query interface with a single search text box, up through advanced interfaces intended for trained archivists, and video data mining applications with visualization. For example, Fig. 10.4 shows the frequency of occurrence of the term "world cup soccer" over a several-year collection of video programs. The visualization interface is interactively generated and the histogram columns are hyperlinked to generate queries restricted to a particular year. Other views of user interfaces built on the MIRACLE platform are shown in Figs. 6.16 – 6.18.

10.7.10 Performance

Table 10.4 gives a sense of the relative complexity of the various content processing operations. The real-time factor is the ratio of processing time to media play time for a particular processing node. Note that these figures do not represent the best achievable values since extensive optimizations such as parallelization have not been undertaken. The indicated indexing operations execute faster than real-time for this particular processing node, but the addition of media transcoding for multiple devices would require an additional node running in parallel for real-time operation. The most time-consuming operation is transcoding for standard definition resolution applications including IPTV, however optimized solutions employing digital signal processing hardware acceleration can perform this in real-time. In fact, encoders for IPTV applications routinely encode HD resolutions as well as lower resolution streams (for picture-in-picture applications) simultaneously.

Fig. 10.4. Mining a multi-year video archive based on content.

Table 10.4. Media processing real-time ratios for a 30 minute 720x480 MPEG-2 file (Dual Intel Pentium Xeon Dual Core 3.0GHz, Dell PE 1950, 15K RPM SAS).

Indexing	**Real-time Ratio**
Video Processing	0.20
Face detection (key frames only)	0.09
CC processing	0.03
Topic segmentation	0.03
Prepare audio	0.09
Speech recognition	0.49
Keyword extraction	0.06
Subtotal	0.99
Transcoding	
MPEG-2 SD to WMV QVGA	0.49
WMV QVGA to MP4 QVGA	0.17
WMV QVGA to WMV QCIF	0.07
WMV QVGA to WMA	0.07
MPEG-2 SD to WMV SD	1.93
Subtotal	2.73
Grand Total	3.72

The processing and transcoding operations can be selected to form an execution profile for a particular application. The full suite represented in Table 10.4 is used for recording episodic broadcast content, where processing servers utilize the idle time between episode broadcasts to generate the transcoded versions. A second profile is used for continuous acquisition from DVRs where streaming proxy resolution is desired. In this case, speech recognition and closed caption alignment is not performed, and only a single QVGA WMV transcoded stream is generated. Another processing system configuration profile for locally indexing content on Media Center PCs does not require any transcoding since the media can be played back from local files.

10.8 Conclusion

The field of content-based video search involves complex image processing algorithms that are often data driven and require verification against test data. Efficient implementations are required to ensure performance on a wide variety of content types, and demonstrate that the approaches are generalizable. Many research groups have undertaken the task of implementing systems and over the years the complexity and capability of these systems continues to grow. We have taken a brief look at a handful of the well known systems, and some relevant but less well known commercial systems. We discussed a practical system in detail to convey an intuition for the computing requirements for certain content processing operations. We described an end-to-end content processing platform suitable for algorithm development and prototyping novel services based on automated metadata augmentation.

References

[Bach06] Bach, J. et al.: Virage image search engine: An open framework for image management, *Storage and Retrieval for Image and Video Databases (SPIE)*, pp. 76–87 (1996).

[Basso00] Basso, A. et al.: PRISM, an IP-based architecture for broadband access to TV and other streaming media, *Proc. 10th Int. Workshop Network and Operating System Support for Digital Audio and Video*, Univ. of North Carolina at Chapel Hill (2000).

[Bender88] Bender, W. and Chesnais, P.: Network Plus, *Proceedings of the SPIE* 900 81–6, Los Angeles, CA (1988).

[Burrelles08] Burrelles*Luce*, Inc. http://www.burrellesluce.com/AboutUs/, cited 9 Feb 2008.

[Chang98] Chang, S.F., Chen, W., and Sundaram, H.: VideoQ: a fully automated video retrieval system using motion sketches, *Applications of Computer Vision,* pp. 270–271 (1998).

[Christel95] Christel, M. et al.: Informedia digital video library, *Communications of the ACM*, **38**(4), pp. 57–58 (1995).

[Christel00] Christel, M. et al.: Interactive maps for a digital video library, *IEEE Multimedia*, **7**(1), pp. 60–67 (2000).

[Critical08] Critical TV Pro, Critical Mention, Inc. http://www.criticalmention.com, cited 8 Feb 2008.

[Davis04] Davis, A. et al.: Retrieving what's relevant in audio and video: statistics and linguistics in combination, Proceedings of RIAO (2004).

[Dimit03] Dimitrova, N., et al.: The Video Scout System: Content-based analysis and retrieval for personal video recorders, in *Handbook of Video Databases*, CRC Press (2003).

[Eber99] Eberman B., et al.: Indexing Multimedia for the Internet. In *Visual Information and Information Systems*. D. P. Huijsmans and Arnold W.M. Smeulders (Eds.) Springer-Verlag (1999).

[FastTalk08] IMTC Projects: Fast-Talk Communications, http://www.imtc.gatech.edu/projects/technology/fasttalk.html, cited 9 Feb 2008.

[Fritz03] Fritz, M., Singingfish: Advancing the art of multimedia search, *EContent*, **26**(4) pp. 52–53 (2003).

[Gibbon99] Gibbon, D. et al.: Browsing and Retrieval of Full Broadcast-Quality Video, Proc. Packet Video Conference (1999).

[Gibbon03] Gibbon, D., et al.: Creating Personalized Video Presentation using Multimodal Processing, *Handbook of Video Databases Design and Applications*, CRC Press, pp. 1107–1132 (2003).

[Gibbon06] Gibbon, D., Liu, Z. and Shahraray, B.: The MIRACLE Video Search Engine, IEEE Consumer Communications and Networking Conference (2006).

[Hanjalic04] Hanjalic, A.: *Content-based Analysis of Digital Video*, Kluwer, Norwell, MA (2004).

[Haputman95] Hauptmann, A. et al.: Speech for Multimedia Information Retrieval, User Interface Software and Technology Conference (1995).

[IBM] IBM technology translates Arabic media broadcasts to English, Press Release 9/14/06. Yorktown Heights, NY (2006).

[Irani93] Irani, M. and Peleg, S.: Motion analysis for image enhancement: resolution, occlusion, and transparency, *J. Visualization, Computers, and Image Recognition*, **4**, pp. 324–335 (1993).

[Irani95] Irani, M.; Anandan, P.; Hsu, S.: Mosaic based representations of video sequences and their applications, *Proceedings of the Fifth International Conference on Computer Vision*, pp. 605–611 (1995).

[Jang99] Jang, P. and Hauptmann, A.: Learning to recognize speech by watching television. *IEEE Intelligent Systems,* **14**(5) pp. 51–58 (1999).

[Jones94] Jones G., et al.: Video Mail Retrieval using Voice: Report on Keyword Definition and Data Collection. University of Cambridge Computer Laboratory, Tech. Report No. 335 (1994).

[Jones95] Jones, G., et al.: Video mail retrieval: the effect of word spotting accuracy on precision, International Conference on Acoustics, Speech, and Signal Processing, pp. 309–312 (1995).

[Lew06] Lew, M. et al.: Content-based multimedia information retrieval: State of the art and challenges. *ACM Trans. Multimedia Computing, Appl.* **2**(1) (2006).

[Liu07] Liu, Z., et al.: AT&T Research at TRECVID 2007. TRECVID 2007 Workshop. (2007).

[Maybury97] Maybury, M., Merlino, A., and Morey, D.: Broadcast News Navigation using Story Segments, *ACM International Multimedia Conference*, pp. 381–391 (1997).

[Maybury04] Maybury, et al., Personalcasting: Tailored Broadcast News, *International Journal of User Modeling and User-Adapted Interaction*, Special Issue on User Modeling and Personalization for TV, Vol. 14, No. 1, pp. 119–144 (2004).

[Mills00] Mills, T. et al.: AT&TV: Broadcast Television and Radio Retrieval, *Proc.* Recherche d'Informations Assistée par Ordinateur; Computer Assisted Information Retrieval (2000).

[Nexedia08] Nexedia, Speech Intelligence Delivered, http://www.nexidia.com, cited 8 Feb 2008.

[Philips05] Phillips, P. et al.: Overview of the Face Recognition Grand Challenge. *Proceedings of the 2005 IEEE Computer Society Conference on Computer Vision and Pattern Recognition,* vol. 1 pp. 947–954 (2005).

[Rehm00] Rehm, E.: Representing Internet streaming media metadata using MPEG-7 multimedia description schemes. *Proceedings of the 2000 ACM Workshops on Multimedia,* pp. 93–98 (2000).

[Rennert03] Rennert, P.: StreamSage Unsupervised ASR-Based Topic Segmentation. TRECVID 2003 - Text REtrieval Conference, Video Track, pp. 17–18 (2003).

[Rose91] Rose, R.C., and Chang, E.I.; Lippmann, R.P., Techniques for information retrieval from voice messages, In *International Conference on Acoustics, Speech, and Signal Processing,* pp. 317–320 (1991).

[Roukos06] Roukos, S. Multilingual Speech and Text Analytics, IBM, September 2006. http://www.research.ibm.com/jam/TALES.pdf accessed 1 Aug 2007.

[Saraclar04] Saraclar, M., Sproat, R.: Lattice-based search for spoken utterance retrieval. In: Human Language Technology Conference (HLT-NAACL), pp. 129–136 (2004).

[Satoh99] Sato, T. et al.: Name-it: Naming and detection faces in news videos, IEEE Multimedia magazine **6**(1), pp, 22–35 (1999).

[Shahraray95] Shahraray B.: Scene Change Detection and Content-based Sampling of Video Sequences, in Digital Video Compression: Algorithms and Technologies, Robert J. Safranek and Arturo A. Rodriguez Editors, Proc. SPIE 2419 (1995).

[Shahrary95a] Shahraray B., and Gibbon D.: Automatic Generation of Pictorial Transcripts of Video Programs, in Multimedia Computing and Networking 1995, Proc. SPIE 2417 (1995).

[Singhal97] Singhal, A., Choi, J., Hindle, D. and Pereira, F.: ATT at TREC-6: SDR track. In Text REtrieval Conference, pp. 227–232 (1997).

[Smith96] Smith, M. and Kanade, T.: Video Skimming and Characterization through the Combination of Image and Language Understanding, Proceedings of the 1998 IEEE International Workshop on Content-Based Access of Image and Video Databases, pp. 61–70 (1998).

[Smith97] Smith, J. and Chang, S.-F.: An Image and Video Search Engine for the World-Wide Web, Proceedings, IS&T/SPIE Symposium on Electronic Imaging: Science and Technology (EI'97) - Storage and Retrieval for Image and Video Databases V, February (1997).

[Syeda00] Syeda-Mahmood, T. et al.: CueVideo: a system for cross-modal search and browse of video databases, IEEE Conf. on Computer Vision and Pattern Recognition, 2000, vol. 2, pp. 786–787 (2000).

[Stuart03] Stuart, A.: SpeechBot: A Search Engine for Sound http://www.hpl.hp.com/news/2003/apr_jun/SpeechBot.html cited 9 Feb 2008.

[Snoek06] Snoek, C., et al.: The Challenge Problem for Automated Detection of 101 Semantic Concepts in Multimedia. Proceedings of ACM Multimedia, pp. 421–430 (2006).

[Strassel03] Stephanie, S. et al.: Shared resources for robust speech-to-text technology, EUROSPEECH, pp. 1609–1612 (2003).

[Teodosio93] Teodosio, L. and Bender, W.: Salient Video Stills: Content and Context Preserved, *Proceedings of ACM Multimedia* (1993).

[VanThong00] Van Thong, J.-M.: SpeechBot: A Speech Recognition based Audio Indexing System for the Web, International Conference on Computer-Assisted Information Retrieval, Recherche d'Informations Assistee par Ordinateur (RIAO2000), pp. 106–115 (2000).

[VMS08] VMS-News Monitoring and Advertising Monitoring Solutions http://www.vidmon.com/, cited 9 Feb 2008.

[Wactlar96] Wactlar, H., Stevens, S., Smith, M., Kanade, T.: Intelligent Access to Digital Video: The Informedia Project, *IEEE Computer*, **29**(5), Digital Library Initiative Special Issue (1996).

[Whittaker02] Whittaker, S., et al.: SCANMail: a voicemail interface that makes speech browsable, readable and searchable. *Proceedings of the SIGCHI Conference on Human Factors in Computing Systems: Changing Our World, Changing Ourselves* CHI '02. ACM Press, pp. 275–282 (2002).

[Worring07] Worring, M. et al.: The MediaMill semantic video search engine. Proceedings of the IEEE International Conference on Acoustics, Speech, and Signal Processing, Hono-lulu, Hawaii, USA (2007).

[Yana07] Yanagawa, A., et al.: Columbia University's Baseline Detectors for 374 LSCOM Semantic Visual Concepts, Columbia University ADVENT Technical Report #222-2006-8 (2007).

[VMS08] VMS, New Monitoring and Advertising Monitoring Solutions, http://www.vidmon.com, cited 8 Feb 2008.

11 Current Trends in Video Search

11.1 Introduction

We've seen that video is ubiquitous on the Web, and that video search sites may focus on a particular aspect such as video sharing, aggregation or original content. Regarding search, sites may have a rich tagging capability or support queries similar to traditional databases while others support full text search of dialogs. Our examination of a wide array of video sources intended for various applications brought to light the range of available metadata that may be available for a given class of video material. Some are rich in standardized metadata, while other content sources such as user-contributed have limited metadata and may rely more on feedback from viewers to enrich the metadata and provide ratings so that others may navigate the content more easily. We've reviewed basic digital video technology with a focus on issues related to video search such as compression, transport, metadata representation, container formats, and media player systems. Our introduction to media processing presented the state of the art in metadata extraction from media with the goal of augmenting available metadata for retrieval applications. We examined processing individual media types including text, audio, and video as well as multimodal techniques. Theses methods are embodied in the research systems which have been improving steadily over the years.

In this chapter, we observe certain trends related to video search. We will try to avoid predicting the future, but by articulating where things were and the current state of the art, we can give the reader a sense of the general direction of constituent technologies as well as social trends. Increasing bandwidth, storage density and computational power are familiar to us given Moore's law and they enable advances such as the cost effective evolution from standard definition to high definition video – even for consumer camcorders.

11.2 Video Production

Today's video distribution systems are digital (e.g. DBS, ATSC, DVB) and we are finally arriving at the cut-over from NTSC analog broadcast over the air to ATSC digital in the US. Across the Media and Entertainment (M+E) industry analog production equipment and media are either gone or on the way out. However, film still maintains a foot-hold in some areas such as on the consumer side in disposable cameras and for archiving movies [Cieply07]. Also, many content owners have large analog tape libraries which are costly to transfer to digital formats.

11.2.1 Metadata Retention

At great expense, broadcasters have made the switch over to HD equipment, and this requires end-to-end upgrades from cameras to play out systems. This newer generation of equipment is often IP enabled and this holds the promise of better metadata preservation. Additional instrumentation and logging from devices and systems such as cameras and file based editing systems is allowing more retention of production metadata using standardized formats such as MXF. This should reduce the need for automated processing to recover the program metadata and structure after distribution, but good practices and interoperability must be proven out for this trend to truly take root.

11.2.2 Multiple Distribution Channels

Rather than producing for a single target such as a half hour broadcast slot, content is created with multiple distribution channels in mind including online and mobile. In many cases a broadcaster may reuse content for multiple services, providing adapted versions as appropriate (e.g. footage may be released with different narration on Discovery Channel vs. Discovery Kids). As another example, many news organizations distribute Podcast versions of their productions, with shorter ad breaks. Production resources such as the studio, staff and on-screen talent, already in place for a primary production, can be efficiently exploited to create short-form content, perhaps with little more incremental cost than an additional script (e.g. Katie

Couric's Notebook is a one minute Podcast, produced daily by NBC news that contains material not used for the nightly news broadcasts.)

11.2.3 Mobisodes and Webisodes

In addition to the reuse or repurposing for multiple distribution channels, there is a new body of content that was never conceived to be long form nor main stream. This includes Podcasters like Rocketboom, vlogs like Brotherhood 2.0 and user generated content sources. As users consume more and more content interactively on the Web, or on the go on portable devices, the trend toward short form episodic content will continue, perhaps offering new paradigms for linking from one short clip to the next in a sequential manner. On the research side, Hart points out the value of searching short form content, which has not been a major research focus – many of the multimedia retrieval evaluations and benchmark datasets focus on long form content (e.g. such tasks as summarization, shot and story segmentation are most useful for long form content) [Hart05].

The Web is a great vehicle for bringing in value from archives of shelved content such as old TV series. The costs are low, but of course the return is not comparable to newly produced content. Also there is a risk of loss of syndication revenue due to audience fatigue, but differing demographics between on-line and TV viewers mitigate this somewhat. In the past, the release of the content on the Web is "last" after broadcast, syndication, DVD, etc. But recently, creators are experimenting with simultaneous release or other strategies to link linear, Web and mobile releases and cross-promote them [Welsh07].

11.3 Video Distribution

MPEG-2 is certainly not going away; there a millions of DVD players and installed set-top boxes that do not support MPEG-4, but newer deployments such as IPTV and mobile devices such as iPods and cell phones use MPEG-4 to provide higher quality at lower bitrates. Combining this with the increased storage capacity of affordable disk drives enables larger volumes of video to be stored at the end user terminal equipment such as DVRs – increasing the need for video search and categorization systems. On the professional production side HDCAM SR brings MPEG-4 (part 2 in this case) into mainstream use at the very start of the content life cycle in addition to distribution.

11.3.1 Streaming Protocols

UDP streaming protocols have the flexibility to be optimized for the application of media delivery. For a given bandwidth, UDP transport provides the best quality – rather than using bandwidth for retransmission of lost packets, transmission errors are concealed at the receiver and more bandwidth is available for the payload. However, using HTTP over TCP provides reliable transport and does not require special firewall configuration; it practically guarantees that if a user can see a Web link to the video, then they will be able to access the video itself. Reliable transport means that extra bits are not used for error resilience, so more efficient coding may be used (e.g. long GoP). The requirement of a large play-out buffer at the client is not a concern given modern clients. The simplification provided by the move to HTTP for media delivery greatly improves reliability, yielding a better quality of experience for the video Web. Deployed Web load balancing and content delivery systems can be reused for cost effective media delivery. Interestingly, for short form content where random access is not critical, HTTP streaming using byte range requests is not necessary, and the situation is essentially that of progressive download which was an early form of Web media delivery.

Media format wars have been on-going since before the Web, so we are unlikely to reach a state where there is a single media format. However, recently Flash video has emerged as the format of choice due to the wide installed base of players [Yan05]. Surprisingly, even Microsoft's UGC site Soapbox uses Flash as it attempts to win viewers from YouTube. For users, this again promotes ease of use for the video Web, and for video search engines, the prospect of Flash ingest as well as Flash distribution may be a requirement. The addition of H.264 support for the Flash Player will increase the adoption on the Web, and to devices in particular [Lynch07].

11.3.2 Electronic Sell Through

The millions of songs sold through iTunes clearly impact physical media (CD) sales, and many efforts are underway to enable distribution of video electronically through download or on-demand services. However, customers still like the visceral experience of opening and owing a disc in their library. They can also be assured that the music from the CD will be compatible with any portable media player that they will ever own. Contrary to this trend, we find the introduction of new physical media formats for HD movies: HD-DVD and BlueRay. In this case, the sheer size of the media files makes electronic distribution cumbersome. So "download to own" may not be as popular as purchasing the disc, but on the other hand, "download to rent" is beginning to make inroads into the video rental market where it can open up the possibility of near instant gratification for users, no late fees or travel to video stores, and vast collections of titles available. What are the implications for online video search? There will be increasingly accessible vast, well organized collections of movie metadata and the corresponding media will be available online with DRM. Yet, services for organizing personal media collections that include some support for removable media will continue to be valuable.

11.3.3 Peer-to-peer Delivery

The traditional client–server media streaming architectures, which have been migrating to HTTP and benefiting from caching improvements provided by content distribution systems, are experiencing competition from peer-to-peer (P2P) distribution technologies. Once feared by content owners as decentralized and therefore difficult to litigate vehicles for piracy, P2P can be used for legitimate content distribution [BBC06]. Originators benefit from vastly decreased distribution costs and users enjoy rapid response for downloading media.

11.3.4 Managed Download

Distribution of content where a download manager takes care of caching the media locally removes any networking variability and produces a quality of experience for the user that is not available with streaming media. There are many examples on the desktop, and perhaps iTunes is the download manager with the largest installed base. This "over the top" (OTT) distribution method for VoD content represents a challenge to existing cable and emerging IPTV VoD systems which rely on guaranteed QoS associated with tightly managed networks. OTT or best effort delivery is seen by some as a legitimate challenge for distribution of linear content as well. The service provider view is that TV viewers will settle for nothing less than instant channel change while watching live HDTV, and telecommunications companies are making huge investments to upgrade network infrastructure to deliver IPTV to support these capabilities. It seems likely that OTT will not fully supplant managed delivery of linear content for the mass market for several years.

11.3.5 Syndication

Users want to view video from all sources from a single interface, even though providers might wish them to visit their portal and see only their content. Single source destinations will not go away, in fact we are seeing that every TV series has its own site for fan interaction, etc., but content is also being syndicated to meet the users' needs. For content discovery, video search engines will use syndication protocols and the value of a blind Web crawl has diminished.

11.4 The Video Web and User Interaction

11.4.1 Web-Based Editing

For UGC, the ability to push raw content directly to Web storage and use Web-based editing tools to create more polished presentations will become a viable alternative, particularly for mobile capture. The notion of content mash-ups, where users pull content from various sources to create novel works, already prevalent today in an offline sense, will become simplified and dynamic. The implications for copyright owners become troublesome, as the "novel" works may be deemed "derivative" by some. For search engines, we can imagine systems that maintain original source content identifiers and edit decision lists in order to efficiently re-use derived index results. This becomes impractical, perhaps, as we move from simple cut edits to more elaborate blending such as dissolves or inserts into regions. Also, it is common for consumers to pull audio from one source and video from another, so identifiers must be managed at a track level.

11.4.2 Media Browsing

The minimalist search box Web form UI was never a good fit for video retrieval, and the lure of promotional revenue has led most video sites to create a dazzling array of enticing content on the landing page. Features such as "most popular" are common on video sites, but less so on text search engines. Also, the notion of "similar" videos displayed after the first query cycle leads the viewer away from subsequent searches and more towards browsing. The fact that video consumption is more lean-back in nature than the "speed reading" mode of operation for text search perhaps justifies the focus on browsing as opposed to search. In many cases, users will visit their favorite video search portal to see "what is on" rather than to start searching based on keywords.

11.4.3 Social Tagging

The author as the single source of content description is yielding to a world where multiple sources of descriptions, both automated via media processing, and manual via social tagging, are common. Search engines must track the source of content descriptions and perhaps apply weights based on authority or consensus when generating ranked results lists.

11.4.4 Dynamic Interfaces

Displaying search results as a one dimensional rank ordered list is the simplest representation. Already, researchers are experimenting with 2-D dynamically rendered views where the search rank dimension is augmented with other attributes such as similarity, or similarity with respect to a particular feature or modality [Worring07].

User interfaces for rendering video search results are dominated by the thumbnail image today, but some sites are adding motion [Blinkx07] to provide users with more insight into the clips' content using the same screen real estate.

Video players embedded in a region of Web page in a browser are predominant and will continue, but new immersive video sites such as Joost, particularly focused on InternetTV applications, which offer full screen replay in full screen with a TV-like paradigm, are just beginning to emerge. This inverts the interface, where traditionally the query and results occupy a large portion of the UI with the video in a small window, to one in which the navigation controls and results occupy a less dominant region of focus for the user. Search engines must work within these confines to render more accurate results in less space – or adopt a modal UI paradigm of shifting back and forth from browse to view modalities, which is disruptive and often can be confusing for users.

As with the immersive, full screen players, video players will become more integrated with the HTML model from a Web developer's perspective. The image tag has served well for integrating still images into Web pages, but attempts to extend this with support for animations and video have been of limited use. Flash authoring offers an alternative, but moves away from the open standards Web interoperability philosophy. HTML 5.0 may include support for video objects within in the document object model [Hickson07].

11.4.5 Video Blogs (vlogs)

Users have several options for publishing their videos; for the highly organized, a video Podcast offers publication for mobility, but constrains creativity and viewer interaction. Video sharing sites are simpler to use and allow for viewer feedback. Posting in blog format is a better fit if textual and video content are both used. The lines of demarcation between these forms are blurred, for example, sharing sites offer the ability to organize content into "channels," and there are tools that allow for re-use of Podcast content in an online interaction setting.

To appease rights owners, automated fingerprinting for content identification to detect copyright infringement is being deployed. There is no substitute for manual review for judging the appropriateness of content for a given audience, however this review may be closed (internal to the service provider) or open to the community to leverage the collective wisdom of viewers. These systems are part of a process that involves auditing and appeals processes since full automation is not practical.

11.4.6 Integrated Collections

To avoid searching in multiple places, users desire breadth of scope, but current revenue models are at odds with this. Beyond comprehensive Web video search, unified retrieval regardless of delivery mechanism (broadcast TV, radio, etc.) will be appreciated by consumers who are more interested in content than how they receive it [Pastra06].

11.5 Television Technology and Consumption

IPTV is in its infancy, particularly in the US and the adoption rate will be a function of business realities as much as technical capabilities. However, IP connectivity to the set-top or flat screen TV itself is arriving from several fronts as big players jockey for control of this highly prized consumer screen; a notable recent entry with an elegant user experience is the AppleTV. Today's gaming consoles are IP enabled and include ample storage for media. Wireless bandwidth is increasing to make in-home connectivity easier, and MOCA and HPNA are in use by service providers for home networking to augment traditional broadcast with IPTV features. These trends open up the possibility for video search on the set-top, however particular focus on the UI will be required for this to be of any use.

11.5.1 Proliferation of Channels

It is not clear how many channels people will need or can tolerate, but since television was invented, the number of channels has only gone up. Service providers advertise the number of channels offered as a selling point, and IPTV promises unlimited channels from international sources to niche content. EPG navigation is currently a problem for users, but this represents an opportunity for video search and content organization systems.

11.5.2 Live to Time Shifted

Cheap hard disc storage with ever increasing density makes DVRs practical and even though HD recording gluttonously consumes storage space, the number of hours of recording available for users is on the rise. Some providers even allow transfer of recordings to removable USB discs (with encryption) for "unlimited" recording. Again, locating content of interest is a potential problem for users but an opportunity for well managed services employing video search technology. It is worth pointing out that there is a limit to the percentage of TV viewing that will move from live to time shifted. Of course live events come to mind, but also scheduled releases of popular programs are likely to be consumed shortly after being broadcast. How else can people discuss them the next day over lunch? Here the DVR may act as a short-term buffer, rather than an archiving device.

11.5.3 Mobile Consumption

It's technically possible to retransmit video received at home to anywhere that IP can reach (e.g. using Sony Location Free TV or Sling Box). However, certain FCC restrictions and sports broadcasting contracts ("blackouts") restrict service providers from offering this to consumers. As online content services mature, not only technically in terms of security and quality, but also in terms of business models where users purchase content licenses for on-line as well as at-home viewing seamlessly, the rationale for slinging the content over to a user's domicile and back again may break down. Also, for personal media collections, users will demand any time, any place access, with back-up security. This suggests an online solution and as the volume of media accessed anytime, anyplace grows, searching and organizing tools will become more valuable.

11.6 Trends in Media Devices

11.6.1 Increased Media Capabilities

Even personal media players that are intended mainly for music are being manufactured with better displays and the ability to play motion video. Still, some consumption scenarios such as while driving an automobile, are not suitable for watching video, but for devices to be more versatile for a range of contexts, video capability is a plus. Considering the wide range of video replay devices from DVD players with SD card slots, connected picture frames, portable gaming devices such as the PSP, and of course iPods, the proliferation of video screens is astounding. Coupling this with ever increasing storage again points to the value for media management systems or personal media library managers to intervene. The more automated these systems become in regards to searching metadata and content itself, the better the user experience will be.

Portable media players are increasing in capability to include Web browsers, and easily connect to wireless networks. Novel interfaces such as multitouch and speech for command and control increase human–machine input bandwidth and ease the adaptation of content to the mobile Web and facilitate video retrieval in the mobile environment. The power of speech for user input is amplified greatly when used in a multimodal context [Johnston07].

The pixel density of mobile device screens is increasing, and entry level digital cameras can now capture VGA resolution video at 30 Hz using inexpensive, fast removable memory. HD consumer video camcorders are rapidly replacing standard definition. These trends result in an increasing abundance of higher quality imagery which must be stored and eventually retrieved.

11.6.2 Increasing Accessibility

The iPod Touch and iPhone are the first iPods to support true closed captioning which will bring video to a wider audience – not only to the hearing impaired, but to public consumption contexts where audio replay is undesirable. The increased number of devices capable of displaying captions will help to motivate content owners to caption more content, and this will generate a good source of index data for search engines.

11.6.3 DRM

DRM hampers the move to digital media as consumers wrestle with incompatible formats and restrictions in usage that they do not experience with CD media. Recently, Amazon is offering DRM-free MP3 songs and the major record labels have agreed to offer content without DRM [Leeds07]. Note that we are yet not seeing a similar trend with respect to high value video content such as movies.

11.6.4 Home Media Systems

High-end consumer media receivers feature IP connectivity for net radio and more affordable dedicated devices have been on the market for some time. LCD TVs such as HP's MediaSmart line now include an RJ-45 input jack and these devices also support WiFi for increased connection convenience. Instead of isolated components we will move to a seamlessly "connected home" although the current landscape is fraught with complexities and incompatibilities for the consumer. When these are resolved, perhaps via adoption of standards such as DLNA, consumers will be able to access, control and search media archives across multiple devices in the home.

11.7 Media Processing Research

With all of these trends in the production, distribution and consumption of media we will see increasing demand for systems to manage video content; to organize, derive associations, relationships, summarize as well as search and browse ever increasing collections of video and audio. The research community is responding to this changing landscape on several fronts, and is bringing more sophisticated analysis methods to bear on these tasks.

While early algorithms attempted to optimize a single approach, the top performing systems for shot boundary detection, high-level feature extraction and recommendation systems [Bell07] employ multiple individual detectors or classifiers, each of which have been highly optimized, and attempt to combine these in optimal ways. Clearly the computational resources required can become enormous and, as has been the case for decades, video processing continues as one of the most computationally demanding of all general purpose computing applications. Thankfully, methods that were impractical a few years ago can be used as a component of multimodal approaches today for increased accuracy. Tasks such as story boundary detection have long been understood to benefit from multimodal processing and this trend continues. As increasing amounts of good training data become available, we may hope to build more generalizable models, and also to consider genre-specific optimizations [Chua04].

Researchers have envisioned that a large number of high level semantic concepts are desirable for video classification and we can trace the progression of classification systems from the early work in identifying a handful of classes such as indoor vs. outdoor up through 39 concepts to 101 and then 394 and beyond [Haupt04, Haupt05].

For video retrieval, as we have seen, the trend is to move from high level metadata, to time varying metadata and to less reliance on textual descriptions in any form. We can classify video search into three types; these are following the historical progression from simple to the more complex, with increasing accuracy and retrieval power. Type one video search is simply based on high level metadata such as title, while type two adds the ability to leverage detailed time-based metadata [Blinkx07a]. Type three involves higher level semantics and can extract features dynamically from multimedia queries [Tseng04]. Many current video classification tasks can be achieved by operating on still frames and there is a new focus on tasks that require true video processing such as motion analysis and event detection.

Along these lines, a recent proposal for TRECVID evaluation involves surveillance event detection. This also brings a new dimension because the data includes multiple video streams capturing a single event. The concept of multistream processing is more general; video conferencing with telepresence using multiple cameras is a valuable area of focus for retrieval system research. Advanced conferencing systems utilize HD displays and cameras. Pan, tilt and zoom are automated and some systems propose the use of omnidirectional cameras [Rui04].

As mentioned, content segmentation is still an area of focus, but with the increasing abundance of short-form content, other tasks involving clip collections in aggregate such as recommendation, association, similarity detection, near-duplicate detection [Zhang04], become more important.

Research trends and advances in the area of content adaptation are addressed in [Chang05] with particular emphasis on summarization, mosaicing, transcoding and standards-based representations. Going forward, the authors point out the need for formal analytical foundations for video content adaptation. In [Chang02], the view of the content chain from end to end is put forth, with several main research directions pointed out in the area of metadata generation from media processing as well as the importance of capturing feedback during retrieval. The notion of reverse engineering of the production process, implicit in many other works but often ignored, is brought to light. Also the impact or value of creating solutions is considered; this point is all too often neglected by researchers when choosing an area of focus. A realistic examination of the applicability of Multimedia Information Retrieval (MIR) was addressed in a panel [Jaimes05] which raised such questions as what is the killer application for MIR and is MIR really necessary given that most search today is not truly content based if we exclude textual content.

Roach et al. provide a good overview of the state of the art in video classification research and provide a taxonomy for organizing tasks such as genre detection and summarization [Roach02]. In 2005, Hauptman summarized ten years of video retrieval work based on contributions form CMU as well as the broader research community [Haupt05]. Image retrieval publication trends have been studied by [Datta04].

The MediaMill project [Worring07] focuses on semantic video search, and has produced several advanced browsers for visualizing search results and allowing users to interact with large results sets efficiently. These systems (e.g. "galaxy browser," "cross browser," "fork browser") take full advantage of available graphics rendering capabilities in order build 3-D interfaces that leverage the rapid visual comprehension skills of users. The systems have performed well in evaluations which provide an indication of the value of these highly interactive, highly capable systems will have in future retrieval systems.

11.8 Deployments

In the last chapter we mentioned some notable video retrieval systems developed by research groups, some of which have led to larger deployments on the Internet which attempt to index significant amounts of published media. We noted SpeechBot as an early technical success in this domain; applying speech recognition for retrieval of large repositories of Web media. More recently Podzinger based on BBN's highly regarded speech technology began indexing podcasts based on speech content and RSS metadata. The site indexed both audio and video podcasts and provided extracted context for query results. Rebranded "EveryZing", the site has this to say regarding ingestion formats: "EveryZing can index, search and reference English and Spanish media (language tag in RSS file must begin with "en" or "es"), that are formatted for audio as MP3 files or for video that are formatted as mp4, mov, m4v, flv, mpg, or mpeg files." [EveryZing07].

BlinkxTV gained attention using speech recognition and other methods such as link context derived from page layout proximity to improve retrieval accuracy. They also have broad coverage of many IP media sources. Additional features such as displaying results in a grid with animation also give Blinkx extra panache and may prove to be valuable for results browsing. Based on technology from Cambridge University, the site reports that 111 patents protect the technology which was developed over 12 years, and that there are over 18M hours of index video [Blinkx07]. In addition to speech processing, any other available metadata, including closed captioning, is used to build the index.

There are hundreds of popular video sharing sites, and of course Google's YouTube is the most well known. (Google has its own video site developed prior to its purchase of YouTube.) With the advent of Podcasting, many sites have sprung up that are focused on indexing syndicated content feeds since the technical barriers to entry are relatively low. These sites enjoy the more organized metadata extracted from RSS or similar syndication formats and the technical work is largely in XML processing, database optimization, UI generation and scaling to handle load. The popularity of various sites is quite dynamic, and many social networking sites are participating in video search both as a content source and as a fabric for posting commentary on clips pasted into user pages. In addition to YouTube, sites such as Dabble, Truevo, Clipblast, Metacafe appear on most lists of popular video sites. In addition to Google, the other established names in the search or portal arena such as Yahoo, AOL, and Microsoft (Live Search, Soapbox) all have video search strategies, some involving partnerships. Most have multiple activities underway focused on sharing and mobile as well as traditional video search.

11.9 Conclusion

Digital video technology has been steadily improving for many years, and recently the cost reductions combined with ease of use for capture and editing has resulted in rapid growth in the amount of video being produced. User generated and enterprise promotional material have mushroomed and augmented traditional video production sources. Distribution to mobile devices and set–tops via IPTV, unheard of a few years ago, is commonplace today. More and more video material, even high quality material, is published on the Web – particularly as more people turn to their laptops as an alternative to watching TV. Further, the Web is extending to an ever increasing array of mobile devices. Therefore Web based video search will be a key enabling technology going forward. Media processing technologies, in addition to metadata handling systems, are available today to help users locate desired content and to navigate through huge amounts of video material for both educational and entertainment applications. These technologies, however, have significant room for improvement through algorithmic innovation as well as through the application of novel engineering techniques to improve the overall efficiency to make indexing large video collections practical. These conditions provide great opportunities for research and development to have a significant impact on people's day to day interaction with video information.

References

[Bell07] Bell, R., Koren, Y. and Volinsky, C.: Chasing $1,000,000: How We Won The Netflix Progress Prize. ASA Statistical and Computing Graphics Newsletter. **18**(2) (2007).

[BBC06] BBC News, BBC moves to file-sharing sites, http://news.bbc.co.uk/nolpda/ukfs_news/hi/newsid_6194000/6194929.stm cited June 16 2008, Dec. 20, 2006.

[Blinkx07] Blinkx, About US – Blnkx, http://www.blinkx.com/about, cited 29 Dec 2007.

[Blinkx07a] Blinkx, Blinkx Video SEO Whitepaper, http://us-store.blinkx.com/images/docs/seo.pdf, cited 29 Dec 2007.

[Chang02] Chang, S.F.: The Holy Grail of content-based media analysis. *IEEE Multimedia Magazine*, **9**(2) pp. 6–10 (2002).

[Chang05] Chang, S.F., and Vetro, A.: Video adaptation: concepts, technologies, and open issues. Special Issue on Advances in Video Coding and Delivery, *Proceedings of IEEE* (2005).

[Chua04] Chua, T., Chang, S., Chaisorn, L., and Hsu, W.: Story boundary detection in large broadcast news video archives: techniques, experience and trends. *Proceedings of the 12th Annual ACM international Conference on Multimedia* (2004).

[Cieply07] Cieply, M.: The Afterlife Is Expensive for Digital Movies, New York Times, December 23, 2007.

[EveryZing07] EveryZing: About EveryZing – FAQ, http://www.everyzing.com/about.jsp?section=faq, cited 30 Dec 2007.

[Hart05] Hart, P. E., Pierson, K., and Hull, J. J.: Refocusing multimedia research on short clips. *IEEE MultiMedia* **12**(3) pp. 8–13 (2005).

[Haupt04] Hauptmann, A.: Towards a large scale concept ontology for broadcast video, Third International Conference on Image and Video Retrieval (CIVR'04), pp. 674–675 (2004).

[Haupt05] Hauptmann, A.: Lessons for the future from a decade of Informedia video analysis research, International Conference on Image and Video Retrieval, Lecture Notes in Computer Science, **3568**, pp. 1-10 (2005).

[Hickson07] Hickson, I., ed.: HTML 5.0 Woprking Draft, W3C HTML Working Group (2007).

[Jaimes05] Jaimes, A., Christel, M., Gilles, S., Sarukkai, R., and Ma, W.: Multimedia information retrieval: what is it, and why isn't anyone using it? In Proceedings of the 7th ACM SIGMM international Workshop on Multimedia Information Retrieval (2005).

[Johnston07] Johnston, M., et al.: A Multimodal Interface for Access to Content in the Home. Proceedings of the Association for Computational Linguistics Annual Conference. pp. 376–383 (2007).

[Leeds07] Leeds, J.: Amazon to Sell Warner Music Minus Copy Protection, The New York Times, December 28, 2007.

[Lynch07] Lynch, K.: Video on the Web, W3C Video on the Web Workshop (2007).

[Pastra06] Pastra K. and Piperidis, S.: Video search: new challenges in the pervasive digital video era, *Journal of Virtual Reality and Broadcasting,* **3**(11) (2006).

[Roach02] Roach, M. et al.: Recent trends in video analysis: a taxonomy of video classification problems, In Proceedings of the International

Conference on Internet and Multimedia Systems and Applications (2002).

[Rui04] Rui, Y. et al.: Automating lecture capture and broadcast: technology and videography, *ACM Multimedia Systems Journal* (Springer), **10** pp. 3-15 (2004).

[Tseng04] Tseng, B. et al.: Using MPEG-7 and MPEG-21 for Personalizing Video, *IEEE MultiMedia*, **11**(1) pp. 42–53 (2004).

[Welsh07] Welsh, J.: "Full immersion" at heart of NBC 360 strategy, Digital Spy, May 14 2007.

[Worring07] Worring, M. et al: The MediaMill semantic video search engine, In *Proceedings of the IEEE International Conference on Acoustics, Speech, and Signal Processing*, (2007).

[Yan05] Yan, J. et al.: Searching for flash movies on the Web: A content and context based framework, *World Wide Web Journal*, September, 2005.

[Zhang04] Zhang D.Q. and Chang, S.F.: Detecting image near-duplicate by stochastic attributed relational graph matching with learning, ACM Conference of Multimedia (2004).

Glossary

AAF – Advanced Authoring Format

ADI – CableLabs® Asset Distribution Interface Specification used for VoD metadata

AMDF – Average magnitude difference function used for pitch calculation

AMG – All Media Guide, provider of metadata services

ANNIE – A Nearly-New Information Extraction System

ASF – Microsoft's Advanced Streaming Format

ASR – Automatic Speech Recognition

ATM – Asynchronous Transfer Mode – a networking technology providing guaranteed quality of service (QoS)

ATSC – Advanced Television Standards Committee

ATIS/IIF – Alliance for Telecommunications Industry Solutions / IPTV Interoperability Forum

BiM – Binary format for MPEG-7 provides high compression of XML representations using the schema definition to remove the syntax redundancy and allows separate source coders to be used for sets of element or attribute values

CGM – Consumer generated media

CID – Content Identifier

CMML – Continuous Media Markup Language

CMS – Content management system

COV – Consumer originated video

CSP – Communications Service Provider

DAM – Digital asset management

DBMS – Database management system

DCT – Discrete cosine transform

DOM – Document Object Model: an interface for accessing HTML and XML in a tree structure; used from languages such as JavaScript (ECMAScript).

DMA – FCC defined metropolitan area for television / radio broadcasting, 210 in the US

DLNA – Digital living network alliance; develops standards for home media device interoperability

DVB – Digital video broadcasting

DVR – Digital video recorder

EMD – earth movers distance

ETSI – European Telecommunications Standards Institute

FAR – Frame aspect ratio

GATE – General Architecture for Text Engineering

Geoblocking – restricting content based on location (blackouts)

GMM – Gaussian mixture model

GoP – Group of pictures

GPS – Global positioning system

GXF – General exchange format

HMM – Hidden Markov model

IAR – Image aspect ratio

K-Space – Knowledge Space

Lemmatize – Convert a word to its root form; a more advanced form of stemming.

LSCOM – Large-Scale Concept Ontology for Multimedia workshop sponsored by the Disruptive Technology Office (DTO)

LVCASR, LVASR, VLVASR – Large Vocabulary Continuous Automatic Speech Recognition, sometimes VLVASR for "Very_." Continuous implies that the input speech waveforms are not segmented and may continue without interruption for many minutes. Also the "C" sometimes references "conversational" connoting a task with multiple speakers and differentiating from "read speech" – a less demanding task since there are fewer disfluencies, and better adherence to rules of grammar, etc.

MFCC – Mel-frequency cepstral coefficients, acoustic features used widely in speech signal processing

MIC – Memory in cassette, an NVRAM chip in a tape cassette for improving access time and storing metadata

MPEG – Moving Picture Experts Group, a working group of ISO/IEC charged with development of video and audio encoding standards.

MPF – Metadata Production Framework; Metadata Editor tools from NHK

NEE – Named entity extraction

NLP / NLU – Natural language processing / understanding refers to the study of computational linguistics typically with the goal of recovering some form of semantics or meaning from textual data

OCR – Optical character recognition

OTT - Over the top: delivery of video on the Internet without guaranteed QoS; may refer to download delivery or services like Joost that deliver a TV like experience a potential threat to Cable and IPTV VoD and service providers.

PAR – Pixel aspect ratio

PDA – Personal digital assistant

POS tagging – Part of speech tagging is the NLP operation of assigning tags to input text to classify words as parts of speech like nouns, verbs, etc.

QbH – Query by humming

QoS – quality of service; as opposed to best effort, QoS is provided by the network and used to guarantee bandwidth for streaming media delivery

SAP – Secondary audio program

SDP – Session description protocol for streaming media initialization, or service delivery platform for providers to deliver media services.

Semantic gap – low level extracted features vs. meaning, understanding

Square pixel – 1:1 PAR

SVG – Scaleable vector graphics

SVM – Support vector machine

TRECVID – TREC video retrieval evaluation, sponsored by the National Institute of Standards and Technology (NIST)

TVML – TV program making language from NHK

UCC – User contributed content

UGC – User generated content

VDF – Virage Data Format

VSF – Video Services Forum

VXML – Vector markup language

WPL – Windows Play List: MS extensions to SMIL that allow for representing queries against media libraries, such as "play all with at least a three star rating"

WSX – SMIL with MS extensions

Zipf's Law – a model with roots in NLP commonly used for describing the long tail phenomenon with regard to the popularity of VoD titles

Index

MIX
Papier aus verantwortungsvollen Quellen
Paper from responsible sources
FSC® C105338

If you have any concerns about our products,
you can contact us on
ProductSafety@springernature.com

In case Publisher is established outside the EU,
the EU authorized representative is:
Springer Nature Customer Service Center GmbH
Europaplatz 3, 69115 Heidelberg, Germany

Printed by Libri Plureos GmbH
in Hamburg, Germany